This is an accurate translation of the Preface by M. Nostradamus to his Prophecies. He wrote it as a letter to his son Cesar.

Preface by M. Nostradamus to His Prophecies

Greetings and happiness to Cesar Nostradamus my son. Your late arrival, Cesar Nostredame, my son, has made me spend much time in constant nightly reflection so that I could communicate with you by letter and leave you this reminder, after my death, for the benefit of all men, of which the divine spirit has vouchsafed me to know by means of astronomy. And since it was the Almighty's will that you were not born here in this region and I do not want to talk of years to come but of the months during which you will struggle to grasp and understand the work I shall be compelled to leave you after my death: assuming that it will not be possible for me to leave you such writing as may be destroyed through the injustice of the age. The key to the hidden prediction which you will inherit will be locked inside my heart.

Also bear in mind that the events here described have not yet come to pass, and that all is ruled and governed by the power of Almighty God, inspiring us not by bacchic frenzy nor by enchantments but by astronomical assurances: predictions have been made through the inspiration of divine will alone and the spirit of prophecy in particular.

On numerous occasions and over a long period of time I have predicted specific events far in advance, attributing all to the workings of divine power and inspiration, together with other fortunate or unfortunate happenings, foreseen in their full unexpectedness, which have already come to pass in various regions of the earth. Yet I have wished to remain silent and abandon my work because of the injustice not only of the present time but also for most of the future. I will not commit to writing.

Since governments, sects and countries will undergo such sweeping changes, diametrically opposed to what now obtains, that were I to relate events to come, those in power now - monarchs, leaders of sects and religions - would find these so different from their own imaginings that they would be led to condemn what later centuries will learn how to see and understand. Bear in mind also Our Saviour's words: "Do not give anything holy to the dogs, nor throw pearls in front of swine lest they trample them with their feet and turn on you and tear you apart." For this reason I withdrew my pen from the paper, because I wished to amplify my statement touching the Vulgar Advent by means of ambiguous and enigmatic comments about future causes, even those closest to us and those I have perceived, so that some human change which may

come to pass shall not unduly scandalize delicate sensibilities. The whole work is thus written in a nebulous rather than plainly prophetic form.
So much so that, "You have hidden these things from the wise and the circumspect, that is from the mighty and the rulers, and you have purified those things for the small and the poor," and through Almighty God's will, revealed unto those prophets with the power to perceive what is distant and thereby to foretell things to come. For nothing can be accomplished without this faculty, whose power and goodness work so strongly in those to whom it is given that, while they contemplate within themselves, these powers are subject to other influences arising from the force of good. This warmth and strength of prophecy invests us with its influence as the sun's rays affect both animate and inanimate entities.

We human beings cannot through our natural consciousness and intelligence know anything of God the Creator's hidden secrets, For it is not for us to know the times or the instants, etc.

So much so that persons of future times may be seen in present ones, because God Almighty has wished to reveal them by means of images, together with various secrets of the future vouchsafed to orthodox astrology, as was the case in the past, so that a measure of power and divination passed through them, the flame of the spirit inspiring them to pronounce upon inspiration both human and divine. God may bring into being divine works, which are absolute; there is another level, that of angelic works; and a third way, that of the evildoers.

But my son, I address you here a little too obscurely. As regards the occult prophecies one is vouchsafed through the subtle spirit of fire, which the understanding sometimes stirs through contemplation of the distant stars as if in vigil, likewise by means of pronouncements, one finds oneself surprised at producing writings without fear of being stricken for such impudent loquacity. The reason is that all this proceeds from the divine power of Almighty God from whom all bounty proceeds.

And so once again, my son, if I have eschewed the word prophet, I do not wish to attribute to myself such lofty title at the present time, for whoever is called a prophet now was once called a seer; since a prophet, my son, is properly speaking one who sees distant things through a natural knowledge of all creatures. And it can happen that the prophet bringing about the perfect light of prophecy may make manifest things both human and divine, because this cannot be done otherwise, given that the effects of predicting the future extend far off into time.

God's mysteries are incomprehensible and the power to influence events is bound up with the great expanse of natural knowledge, having its nearest most immediate origin in free will and describing future events which cannot be understood simply through being revealed. Neither can they be grasped through men's interpretations nor through another mode of cognizance or occult power under the firmament, neither in the present nor in the total eternity to come But bringing about such an indivisible eternity through Herculean efforts, things are revealed by the planetary movements.

I am not saying, my son - mark me well, here - that knowledge of such things cannot be implanted in your deficient mind, or that events in the distant future may not be within the understanding of any reasoning being. Nevertheless, if these things current or distant are brought to the awareness of this reasoning and intelligent being they will be neither too obscure nor too clearly revealed.

Perfect knowledge of such things cannot be acquired without divine inspiration, given that all prophetic inspiration derives its initial origin from God Almighty, then from chance and nature. Since all these portents are produced impartially, prophecy comes to pass partly as predicted. For understanding created by the intellect cannot be acquired by means of the occult, only by the aid of the zodiac, bringing forth that small flame by whose light part of the future may be discerned.

Also, my son, I beseech you not to exercise your mind upon such reveries and vanities as drain the body and incur the soul's perdition, and which trouble our feeble frames. Above all avoid the vanity of that most execrable magic formerly reproved by the Holy Scriptures - only excepting the use of official astrology.

For by the latter, with the help of inspiration and divine revelation, and continual calculations, I have set down my prophecies in writing. Fearing lest this occult philosophy be condemned, I did not therefore wish to make known its dire import; also fearful that several books which had lain hidden for long centuries might be discovered, and of what might become of them, after reading them I presented them to Vulcan. And while he devoured them, the flame licking the air gave out such an unexpected light, clearer than that of an ordinary flame and resembling fire from some flashing cataclysm, and suddenly illumined the house as if it were caught in a furnace. Which is why I reduced them to ashes then, so that none might be tempted to use occult labours in searching for the perfect transmutation, whether lunar or solar, of incorruptible metals.

But as to that discernment which can be achieved by the aid of planetary scrutiny, I should like to tell you this. Eschewing any fantastic imaginings, you may through good judgement have insight into the future if you keep to the specific names of places that accord with planetary configurations, and with inspiration places and aspects yield up hidden properties, namely that power in whose presence the three times are understood as Eternity whose unfolding contains them all: for all things are naked and open.

That is why, my son, you can easily, despite your young brain, understand that events can be foretold naturally by the heavenly bodies and by the spirit of prophecy: I do not wish to ascribe to myself the title and role of prophet, but emphasize inspiration revealed to a mortal man whose perception is no further from heaven than the feet are from the earth. I cannot fail, err or be deceived, although I may be as great a sinner as anyone else upon this earth and subject to all human afflictions.

But after being surprised sometimes by day while in a trance, and having long fallen into the habit of agreeable nocturnal studies, I have composed books of prophecies, each containing one hundred astronomical quatrains, which I want to condense somewhat obscurely. The work comprises prophecies from today to the year 3797.

This may perturb some, when they see such a long timespan, and this will occur and be understood in all the fullness of the Republic; these things will be universally understood upon earth, my son. If you live the normal lifetime of man you will know upon your own soil, under your native sky, how future events are to turn out.

For only Eternal God knows the eternity of His light which proceeds from Him, and I speak frankly to those to whom His immeasurable, immense and incomprehensible greatness has been disposed to grant revelations through long, melancholy inspiration, that with the aid of this hidden element manifested by God, there are two principal factors which make up the prophet's intelligence.

The first is when the supernatural light fills and illuminates the person who predicts by astral science, while the second allows him to prophesy through inspired revelation, which is only a part of the divine eternity, whereby the prophet comes to assess what his divinatory power has given him through the grace of God and by a natural gift, namely, that what is foretold is true and ethereal in origin.

And such a light and small flame is of great efficacy and scope, and nothing less than the clarity of nature itself. The light of human nature makes the philosophers so sure of themselves that with the principles of the first cause they reach the loftiest doctrines and the deepest abysses.

But my son, lest I venture too far for your future perception, be aware that men of letters shall make grand and usually boastful claims about the way I interpreted the world, before the worldwide conflagration which is to bring so many catastrophes and such revolutions that scarcely any lands will not be covered by water, and this will last until all has perished save history and geography themselves. This is why, before and after these revolutions in various countries, the rains will be so diminished and such abundance of fire and fiery missiles shall fall from the heavens that nothing shall escape the holocaust. And this will occur before the last conflagration.

For before war ends the century and in its final stages it will hold the century under its sway. Some countries will be in the grip of revolution for several years, and others ruined for a still longer period. And now that we are in a republican era, with Almighty God's aid, and before completing its full cycle, the monarchy will return, then the Golden Age. For according to the celestial signs, the Golden Age shall return, and after all calculations, with the world near to an all-encompassing revolution - from the time of writing 177 years 3 months 11 days - plague, long famine and wars, and still more floods from now until the stated time. Before and after these, humanity shall several times be so severely diminished that scarcely anyone shall be found who wishes to take over the fields, which shall become free where they had previously been tied.

This will be after the visible judgement of heaven, before we reach the millennium which shall complete all. In the firmament of the eighth sphere, a dimension whereon Almighty God will complete the revolution, and where the constellations will resume their motion which will render the earth stable and firm, but only if He will remain unchanged for ever until His will be done.

This is in spite of all the ambiguous opinions surpassing all natural reason, expressed by Mahomet; which is why God the Creator, through the ministry of his fiery agents with their flames, will come to propose to our perceptions as well as our eyes the reasons for future predictions.

Signs of events to come must be manifested to whomever prophesies. For prophecy which stems from exterior illumination is part of that light and seeks to ally with it and bring it into being so that the part which seems to possess the faculty of understanding is not subject to a sickness of the mind.

Reason is only too evident. Everything is predicted by divine afflatus and thanks to an angelic spirit inspiring the one prophesying, consecrating his predictions through divine unction. It also divests him of all fantasies by means of various nocturnal apparitions, while with daily certainty he prophesies through the science of astronomy, with the aid of sacred prophecy, his only consideration being his courage in freedom.

So come, my son, strive to understand what I have found out through my calculations which accord with revealed inspiration, because now the sword of death approaches us, with pestilence and war more horrible than there has ever been - because of three men's work - and famine. And this sword shall smite the earth and return to it often, for the stars confirm this upheaval and it is also written: "I shall punish their injustices with iron rods, and shall strike them with blows."

For God's mercy will be poured forth only for a certain time, my son, until the majority of my prophecies are fulfilled and this fulfillment is complete. Then several times in the course of the doleful tempests the Lord shall say: Therefore I shall crush and destroy and show no mercy; and many other circumstances shall result from floods and continual rain of which I have written more fully in my other prophecies, composed at some length, not in a chronological sequence, in prose, limiting the places and times and exact dates so that future generations will see, while experiencing these inevitable events, how I have listed others in clearer language, so that despite their obscurities these things shall be understood: When the time comes for the removal of ignorance, the matter will be clearer still.

So in conclusion, my son, take this gift from your father M. Nostradamus, who hopes you will understand each prophecy in every quatrain herein. May Immortal God grant you a long life of good and prosperous happiness.

Salon, 1 March 1555

Century I

This is the first Century by Nostradamus, first printed on May 4, 1555 in Lyons by Bonhomme. This first edition contained the Preface to his son Cesar and 353 quatrains. A reprint was done by Bareste in 1840; unfortunately, the original was lost.

[Note: This is not my translation. It was found on the Internet a few years ago on a site that no longer exists. I have cleaned up gross misspellings and typos but have otherwise left the text intact. At some point, I hope to update with what I consider better translation. Due to lack of time on my part, this will have to suffice for now.

For the French version, go to Nostradamiana at:
http://www.astrologer.ru:8003/Nostradamiana/centuries-eng.html.]

CENTURY I

1
Sitting alone at night in secret study;
it is placed on the brass tripod.
A slight flame comes out of the emptiness and

makes successful that which should not be believed in vain.

2
The wand in the hand is placed in the middle of the tripod's legs.
With water he sprinkles both the hem of his garment and his foot.
A voice, fear: he trembles in his robes.
Divine splendor; the God sits nearby.

3
When the litters are overturned by the whirlwind
and faces are covered by cloaks,
the new republic will be troubled by its people.
At this time the reds and the whites will rule wrongly.

4
In the world there will be made a king
who will have little peace and a short life.
At this time the ship of the Papacy will be lost,
governed to its greatest detriment.

5
They will be driven away for a long drawn out fight.
The countryside will be most grievously troubled.
Town and country will have greater struggle.
Carcassonne and Narbonne will have their hearts tried.

6
The eye of Ravenna will be forsaken,
when his wings will fail at his feet.
The two of Bresse will have made a constitution
for Turin and Vercelli, which the French will trample underfoot

7
Arrived too late, the act has been done.
The wind was against them, letters intercepted on their way.
The conspirators were fourteen of a party.
By Rousseau shall these enterprises be undertaken.

8
How often will you be captured, O city of the sun ?
Changing laws that are barbaric and vain.
Bad times approach you. No longer will you be enslaved.
Great Hadrie will revive your veins.

9
From the Orient will come the African heart

to trouble Hadrie and the heirs of Romulus.
Accompanied by the Libyan fleet
the temples of Malta and nearby islands shall be deserted.

10
A coffin is put into the vault of iron,
where seven children of the king are held.
The ancestors and forebears will come forth from the depths of hell,
lamenting to see thus dead the fruit of their line.

11
The motion of senses, heart, feet and hands
will be in agreement between Naples, Lyon and Sicily.
Swords fire, floods, then the noble Romans drowned,
killed or dead because of a weak brain.

12
There will soon be talk of a treacherous man, who rules a short time,
quickly raised from low to high estate.
He will suddenly turn disloyal and volatile.
This man will govern Verona.

13
Through anger and internal hatreds, the exiles
will hatch a great plot against the king.
Secretly they will place enemies as a threat,
and his own old (adherents) will find sedition against them.

14
From the enslaved populace, songs, chants and demands,
while Princes and Lords are held captive in prisons.
These will in the future by headless idiots
be received as divine prayers

15
.Mars threatens us with the force of war
and will cause blood to be spilt seventy times.
The clergy will be both exalted and reviled moreover,
by those who wish to learn nothing of them.

16
A scythe joined with a pond in Sagittarius
at its highest ascendant.
Plague, famine, death from military hands;
the century approaches its renewal.

17
For forty years the rainbow will not be seen.
For forty years it will be seen every day.
The dry earth will grow more parched,
and there will be great floods when it is seen.

18
Because of French discord and negligence
an opening shall be given to the Mohammedans.
The land and sea of Siena will be soaked in blood,
and the port of Marseilles covered with ships and sails.

19
When the snakes surround the altar,
and the Trojan blood is troubled by the Spanish.
Because of them, a great number will be lessened.
The leader flees, hidden in the swampy marshes.

20
The cities of Tours, Orleans, Blois, Angers, Reims and Nantes
are troubled by sudden change.
Tents will be pitched by (people) of foreign tongues;
rivers, darts at Rennes, shaking of land and sea.

21
The rock holds in its depths white clay
which will come out milk-white from a cleft
Needlessly troubled people will not dare touch it,
unaware that the foundation of the earth is of clay.

22
A thing existing without any senses
will cause its own end to happen through artifice.
At Autun, Chalan, Langres and the two Sens
there will be great damage from hail and ice.

23
In the third month, at sunrise,
the Boar and the Leopard meet on the battlefield.
The fatigued Leopard looks up to heaven
and sees an eagle playing around the sun.

24
At the New City he is thoughtful to condemn;
the bird of prey offers himself to the Gods.
After victory he pardons his captives.

At Cremona and Mantua great hardships will be suffered.

25
The lost thing is discovered, hidden for many centuries.
Pasteur will be celebrated almost as a God-like figure.
This is when the moon completes her great cycle,
but by other rumors he shall be dishonored.

26
The great man will be struck down in the day by a thunderbolt.
An evil deed, foretold by the bearer of a petition.
According to the prediction another falls at night time.
Conflict at Reims, London, and pestilence in Tuscany.

27
Beneath the oak tree of Gienne, struck by lightning,
the treasure is hidden not far from there.
That which for many centuries had been gathered,
when found, a man will die, his eye pierced by a spring.

28
Tobruk will fear the barbarian fleet for a time,
then much later the Western fleet.
Cattle, people, possessions, all will be quite lost.
What a deadly combat in Taurus and Libra.

29
When the fish that travels over both land and sea
is cast up on to the shore by a great wave,
its shape foreign, smooth and frightful.
From the sea the enemies soon reach the walls.

30
Because of the storm at sea the foreign ship
will approach an unknown port.
Notwithstanding the signs of the palm branches,
afterwards there is death and pillage. Good advice comes too late.

31
The wars in France will last for so many years
beyond the reign of the Castulon kings.
An uncertain victory will crown three great ones,
the Eagle, the Cock, the Moon, the Lion, the Sun in its house.

32
The great Empire will soon be exchanged

for a small place, which soon will begin to grow.
A small place of tiny area
in the middle of which he will come to lay down his scepter.

33
Near a great bridge near a spacious plain
the great lion with the Imperial forces
will cause a falling outside the austere city.
Through fear the gates will be unlocked for him.

34
The bird of prey flying to the left,
before battle is joined with the French, he makes preparations.
Some will regard him as good, others bad or uncertain.
The weaker party will regard him as a good omen.

35
The young lion will overcome the older one,
in a field of combat in single fight:
He will pierce his eyes in their golden cage;
two wounds in one, then he dies a cruel death.

36
Too late the king will repent
that he did not put his adversary to death.
But he will soon come to agree to far greater things
which will cause all his line to die.

37
Shortly before sun set, battle is engaged.
A great nation is uncertain.
Overcome, the sea port makes no answer,
the bridge and the grave both in foreign places.

38
The Sun and the Eagle will appear to the victor.
An empty answer assured to the defeated.
Neither bugle nor shouts will stop the soldiers.
Liberty and peace, if achieved in time through death.

39
At night the last one will be strangled in his bed
because he became too involved with the blond heir elect.
The Empire is enslaved and three men substituted.
He is put to death with neither letter nor packet read.

40
The false trumpet concealing madness
will cause Byzantium to change its laws.
From Egypt there will go forth a man who wants
the edict withdrawn, changing money and standards.

41
The city is besieged and assaulted by night;
few have escaped; a battle not far from the sea.
A woman faints with joy at the return of her son,
poison in the folds of the hidden letters.

42
The tenth day of the April Calends, calculated in Gothic fashion
is revived again by wicked people.
The fire is put out and the diabolic gathering
seek the bones of the demon of Psellus.

43
Before the Empire changes
a very wonderful event will take place.
The field moved, the pillar of porphyry
put in place, changed on the gnarled rock.

44
In a short time sacrifices will be resumed,
those opposed will be put (to death) like martyrs.
The will no longer be monks, abbots or novices.
Honey shall be far more expensive than wax.

45
A founder of sects, much trouble for the accuser:
A beast in the theater prepares the scene and plot.
The author ennobled by acts of older times;
the world is confused by schismatic sects.

46
Very near Auch, Lectoure and Mirande
a great fire will fall from the sky for three nights.
The cause will appear both stupefying and marvelous;
shortly afterwards there will be an earthquake.

47
The speeches of Lake Leman will become angered,
the days will drag out into weeks,
then months, then years, then all will fail.

The authorities will condemn their useless powers.

48
When twenty years of the Moon's reign have passed
another will take up his reign for seven thousand years.
When the exhausted Sun takes up his cycle
then my prophecy and threats will be accomplished.

49
Long before these happenings
the people of the East, influenced by the Moon,
in the year 1700 will cause many to be carried away,
and will almost subdue the Northern area.

50
From the three water signs will be born a man
who will celebrate Thursday as his holiday.
His renown, praise, rule and power will grow
on land and sea, bringing trouble to the East.

51
The head of Aries, Jupiter and Saturn.
Eternal God, what changes !
Then the bad times will return again after a long century;
what turmoil in France and Italy.

52
Two evil influences in conjunction in Scorpio.
The great lord is murdered in his room.
A newly appointed king persecutes the Church,
the lower (parts of) Europe and in the North.

53
Alas, how we will see a great nation sorely troubled
and the holy law in utter ruin.
Christianity (governed) throughout by other laws,
when a new source of gold and silver is discovered.

54
Two revolutions will be caused by the evil scythe bearer
making a change of reign and centuries.
The mobile sign thus moves into its house:
Equal in favor to both sides.

55
In the land with a climate opposite to Babylon

there will be great shedding of blood.
Heaven will seem unjust both on land and sea and in the air.
Sects, famine, kingdoms, plagues, confusion.

56
Sooner and later you will see great changes made,
dreadful horrors and vengeances.
For as the moon is thus led by its angel
the heavens draw near to the Balance.

57
The trumpet shakes with great discord.
An agreement broken: lifting the face to heaven:
the bloody mouth will swim with blood;
the face anointed with milk and honey lies on the ground.

58
Through a slit in the belly a creature will be born with two heads
and four arms: it will survive for some few years.
The day that Alquiloie celebrates his festivals
Fossana, Turin and the ruler of Ferrara will follow.

59
The exiles deported to the islands
at the advent of an even more cruel king
will be murdered. Two will be burnt
who were not sparing in their speech.

60
An Emperor will be born near Italy,
who will cost the Empire very dearly.
They will say, when they see his allies,
that he is less a prince than a butcher.

61
The wretched, unfortunate republic
will again be ruined by a new authority.
The great amount of ill will accumulated in exile
will make the Swiss break their important agreement.

62
Alas! what a great loss there will be to learning
before the cycle of the Moon is completed.
Fire, great floods, by more ignorant rulers;
how long the centuries until it is seen to be restored.

63

Pestilences extinguished, the world becomes smaller,
for a long time the lands will be inhabited peacefully.
People will travel safely through the sky (over) land and seas:
then wars will start up again.

64

At night they will think they have seen the sun,
when the see the half pig man:
Noise, screams, battles seen fought in the skies.
The brute beasts will be heard to speak.

65

A child without hands, never so great a thunderbolt seen,
the royal child wounded at a game of tennis.
At the well lightning strikes, joining together
three trussed up in the middle under the oaks.

66

He who then carries the news,
after a short while will (stop) to breathe:
Viviers, Tournon, Montferrand and Praddelles;
hail and storms will make them grieve.

67

The great famine which I sense approaching
will often turn (in various areas) then become worldwide.
It will be so vast and long lasting that (they) will grab
roots from the trees and children from the breast.

68

O to what a dreadful and wretched torment
are three innocent people going to be delivered.
Poison suggested, badly guarded, betrayal.
Delivered up to horror by drunken executioners.

69

The great mountain, seven stadia round,
after peace, war, famine, flooding.
It will spread far, drowning great countries,
even antiquities and their mighty foundations.

70

Rain, famine and war will not cease in Persia;
too great a faith will betray the monarch.
Those (actions) started in France will end there,

a secret sign for on to be sparing.

71
The marine tower will be captured and retaken three times
by Spaniards, Barbarians and Ligurians.
Marseilles and Aix, Ales by men of Pisa,
devastation, fire, sword, pillage at Avignon by the Turinese.

72
The inhabitants of Marseilles completely changed,
fleeing and pursued as far as Lyons.
Narbonne, Toulouse angered by Bordeaux;
the killed and captive are almost one million.

73
France shall be accused of neglect by her five partners.
Tunis, Algiers stirred up by the Persians.
Leon, Seville and Barcelona having failed,
they will not have the fleet because of the Venetians.

74
After a rest they will travel to Epirus,
great help coming from around Antioch.
The curly haired king will strive greatly for the Empire,
the brazen beard will be roasted on a spit.

75
The tyrant of Siena will occupy Savona,
having won the fort he will restrain the marine fleet.
Two armies under the standard of Ancona:
the leader will examine them in fear.

76
The man will be called by a barbaric name
that three sisters will receive from destiny.
He will speak then to a great people in words and deeds,
more than any other man will have fame and renown.

77
A promontory stands between two seas:
A man who will die later by the bit of a horse;
Neptune unfurls a black sail for his man;
the fleet near Gibraltar and Rocheval.

78
To an old leader will be born an idiot heir,

weak both in knowledge and in war.
The leader of France is feared by his sister,
battlefields divided, conceded to the soldiers.

79
Bazas, Lectoure, Condom, Auch and Agen
are troubled by laws, disputes and monopolies.
Carcassone, Bordeaux, Toulouse and Bayonne will be ruined
when they wish to renew the massacre.

80
From the sixth bright celestial light
it will come to thunder very strongly in Burgundy.
Then a monster will be born of a very hideous beast:
In March, April, May and June great wounding and worrying.

81
Nine will be set apart from the human flock,
separated from judgment and advise.
Their fate is to be divided as they depart.
K. Th. L. dead, banished and scattered.

82
When the great wooden columns tremble
in the south wind, covered with blood.
Such a great assembly then pours forth
that Vienna and the land of Austria will tremble.

83
The alien nation will divide the spoils.
Saturn in dreadful aspect in Mars.
Dreadful and foreign to the Tuscans and Latins,
Greeks who will wish to strike.

84
The moon is obscured in deep gloom,
his brother becomes bright red in color.
The great one hidden for a long time in the shadows
will hold the blade in the bloody wound.

85
The king is troubled by the queen's reply.
Ambassadors will fear for their lives.
The greater of his brothers will doubly disguise his action,
two of them will die through anger, hatred and envy.

86
When the great queen sees herself conquered,
she will show an excess of masculine courage.
Naked, on horseback, she will pass over the river
pursued by the sword: she will have outraged her faith

87
Earthshaking fire from the center of the earth
will cause tremors around the New City.
Two great rocks will war for a long time,
then Arethusa will redden a new river.

88
The divine wrath overtakes the great Prince,
a short while before he will marry.
Both supporters and credit will suddenly diminish.
Counsel, he will die because of the shaven heads.

89
Those of Lerida will be in the Moselle,
kill all those from the Loire and Seine.
The seaside track will come near the high valley,
when the Spanish open every route.

90
Bordeaux and Poitiers at the sound of the bell
will go with a great fleet as fast as Langon.
A great rage will surge up against the French,
when a hideous monster is born near Orgon.

91
The Gods will make it appear to mankind
that they are the authors of a great war.
Before the sky was seen to bee free of weapons and rockets:
the greatest damage will be inflicted on the left.

92
Under one man peace will be proclaimed everywhere,
but not long after will be looting and rebellion.
Because of a refusal, town, land and see will be broached.
About a third of a million dead or captured.

93
The Italian lands near the mountains will tremble.
The Cock and the Lion not strongly united.
In place of fear they will help each other.

Freedom alone moderates the French.

94
The tyrant Selim will be put to death at the harbor
but Liberty will not be regained, however.
A new war arises from vengeance and remorse.
A lady is honored through force of terror.

95
In front of a monastery will be found a twin infant
from the illustrious and ancient line of a monk.
His fame, renown and power through sects and speech
is such that they will say the living twin is deservedly chosen.

96
A man will be charged with the destruction
of temples and sects, altered by fantasy.
He will harm the rocks rather than the living,
ears filled with ornate speeches.

97
That which neither weapon nor flame could accomplish
will be achieved by a sweet speaking tongue in council.
Sleeping, in a dream, the king will see
the enemy not in war or of military blood.

98
The leader who will conduct great numbers of people
far from their skies, to foreign customs and language.
Five thousand will die in Crete and Thessaly,
the leader fleeing in a sea going supply ship.

99
The great king will join
with two kings, united in friendship.
How the great household will sigh:
around Narbon what pity for the children.

100
For a long time a gray bird will be seen in the sky
near Dôle and the lands of Tuscany.
He holds a flowering branch in his beak,
but he dies too soon and the war ends.

Century II

This is the second Century by Nostradamus. It was first published in 1555.

[Note: This is not my translation. It was found on the Internet a few years ago on a site that no longer exists. I have cleaned up gross misspellings and typos but have otherwise left the text intact. At some point, I hope to update with what I consider better translation. Due to lack of time on my part, this will have to suffice for now.

For the French version, go to Nostradamiana at:
http://www.astrologer.ru:8003/Nostradamiana/centuries-eng.html.]

CENTURY II

1
Towards Aquitaine by the British Isles
By these themselves great incursions.
Rains, frosts will make the soil uneven,
Port Selyn will make mighty invasions

2
The blue head will inflict upon the white head
As much evil as France has done them good:
Dead at the sail-yard the great one hung on the branch.
When seized by his own the King will say how much.

3
Because of the solar heat on the sea
From Negrepont the fishes half cooked:
The inhabitants will come to cut them,
When food will fail in Rhodes and Genoa.

4
From Monaco to near Sicily
The entire coast will remain desolated:
There will remain there no suburb, city or town
Not pillaged and robbed by the Barbarians.

5
That which is enclosed in iron and letter in a fish,
Out will go one who will then make war,
He will have his fleet well rowed by sea,
Appearing near Latin land.

6

Near the gates and within two cities
There will be two scourges the like of which was never seen,
Famine within plague, people put out by steel,
Crying to the great immortal God for relief.

7

Amongst several transported to the isles,
One to be born with two teeth in his mouth
They will die of famine the trees stripped,
For them a new King issues a new edict.

8

Temples consecrated in the original Roman manner,
They will reject the excess foundations,
Taking their first and humane laws,
Chasing, though not entirely, the cult of saints.

9

Nine years the lean one will hold the realm in peace,
Then he will fall into a very bloody thirst:
Because of him a great people will die without faith and law
Killed by one far more good-natured.

10

Before long all will be set in order,
We will expect a very sinister century,
The state of the masked and solitary ones much changed,
Few will be found who want to be in their place.

11

The nearest son of the elder will attain
Very great height as far as the realm of the privileged:
Everyone will fear his fierce glory,
But his children will be thrown out of the realm.

12

Eyes closed, opened by antique fantasy,
The garb of the monks they will be put to naught:
The great monarch will chastise their frenzy,
Ravishing the treasure in front of the temples.

13

The body without soul no longer to be sacrificed:
Day of death put for birthday:
The divine spirit will make the soul happy,
Seeing the word in its eternity.

14

At Tours, Gien, guarded, eyes will be searching,
Discovering from afar her serene Highness:
She and her suite will enter the port,
Combat, thrust, sovereign power.

15

Shortly before the monarch is assassinated,
Castor and Pollux in the ship, bearded star:
The public treasure emptied by land and sea,
Pisa, Asti, Ferrara, Turin land under interdict.

16

Naples, Palermo, Sicily, Syracuse,
New tyrants, celestial lightning fires:
Force from London, Ghent, Brussels and Susa,
Great slaughter, triumph leads to festivities.

17

The field of the temple of the vestal virgin,
Not far from Elne and the Pyrenees mountains:
The great tube is hidden in the trunk.
To the north rivers overflown and vines battered.

18

New, impetuous and sudden rain
Will suddenly halt two armies.
Celestial stone, fires make the sea stony,
The death of seven by land and sea sudden.

19

Newcomers, place built without defense,
Place occupied then uninhabitable:
Meadows, houses, fields, towns to take at pleasure,
Famine, plague, war, extensive land arable.

20

Brothers and sisters captive in diverse places
Will find themselves passing near the monarch:
Contemplating them his branches attentive,
Displeasing to see the marks on chin, forehead and nose.

21

The ambassador sent by biremes,
Halfway repelled by unknown ones:

Reinforced with salt four triremes will come,
In Euboea bound with ropes and chains.

22
The imprudent army of Europe will depart,
Collecting itself near the submerged isle:
The weak fleet will bend the phalanx,
At the navel of the world a greater voice substituted.

23
Palace birds, chased out by a bird,
Very soon after the prince has arrived:
Although the enemy is repelled beyond the river,
Outside seized the trick upheld by the bird.

24
Beasts ferocious from hunger will swim across rivers:
The greater part of the region will be against the Hister,
The great one will cause it to be dragged in an iron cage,
When the German child will observe nothing.

25
The foreign guard will betray the fortress,
Hope and shadow of a higher marriage:
Guard deceived, fort seized in the press,
Loire, Saone, Rhone, Garonne, mortal outrage.

26
Because of the favor that the city will show
To the great one who will soon lose the field of battle,
Fleeing the Po position, the Ticino will overflow
With blood, fires, deaths, drowned by the long-edged blow.

27
The divine word will be struck from the sky,
One who cannot proceed any further:
The secret closed up with the revelation,
Such that they will march over and ahead.

28
The penultimate of the surname of the Prophet
Will take Diana [Thursday] for his day and rest:
He will wander far because of a frantic head,
And delivering a great people from subjection.

29

The Easterner will leave his seat,
To pass the Apennine mountains to see Gaul:
He will transpire the sky, the waters and the snow,
And everyone will be struck with his rod.

30
One who the infernal gods of Hannibal
Will cause to be reborn, terror of mankind
Never more horror nor worse of days
In the past than will come to the Romans through Babel.

31
In Campania the Capuan [river] will do so much
That one will see only fields covered by waters:
Before and after the long rain
One will see nothing green except the trees.

32
Milk, frog's blood prepared in Dalmatia.
Conflict given, plague near Treglia:
A great cry will sound through all Slavonia,
Then a monster will be born near and within Ravenna.

33
Through the torrent which descends from Verona
Its entry will then be guided to the Po,
A great wreck, and no less in the Garonne,
When those of Genoa march against their country.

34
The senseless ire of the furious combat
Will cause steel to be flashed at the table by brothers:
To part them death, wound, and curiously,
The proud duel will come to harm France.

35
The fire by night will take hold in two lodgings,
Several within suffocated and roasted.
It will happen near two rivers as one:
Sun, Sagittarius and Capricorn all will be reduced.

36
The letters of the great Prophet will be seized,
They will come to fall into the hands of the tyrant:
His enterprise will be to deceive his King,
But his extortions will very soon trouble him.

37
Of that great number that one will send
To relieve those besieged in the fort,
Plague and famine will devour them all,
Except seventy who will be destroyed.

38
A great number will be condemned
When the monarchs will be reconciled:
But for one of them such a bad impediment will arise
That they will be joined together but loosely.

39.
One year before the Italian conflict,
Germans, Gauls, Spaniards for the fort:
The republican schoolhouse will fall,
There, except for a few, they will be choked dead.

40
Shortly afterwards, without a very long interval,
By sea and land a great uproar will be raised:
Naval battle will be very much greater,
Fires, animals, those who will cause greater insult.

41
The great star will burn for seven days,
The cloud will cause two suns to appear:
The big mastiff will howl all night
When the great pontiff will change country.

42
Cock, dogs and cats will be satiated with blood
And from the wound of the tyrant found dead,
At the bed of another legs and arms broken,
He who was not afraid to die a cruel death.

43
During the appearance of the bearded star.
The three great princes will be made enemies:
Struck from the sky, peace earth quaking,
Po, Tiber overflowing, serpent placed upon the shore.

44
The Eagle driven back around the tents
Will be chased from there by other birds:

When the noise of cymbals, trumpets and bells
Will restore the senses of the senseless lady.

45.
Too much the heavens weep for the Androgyne begotten,
Near the heavens human blood shed:
Because of death too late a great people re-created,
Late and soon the awaited relief comes.

46
After great trouble for humanity, a greater one is prepared
The Great Mover renews the ages:
Rain, blood, milk, famine, steel and plague,
Is the heavens fire seen, a long spark running.

47
The great old enemy mourning dies of poison,
The sovereigns subjugated in infinite numbers:
Stones raining, hidden under the fleece,
Through death articles are cited in vain.

48
The great force which will pass the mountains.
Saturn in Sagittarius Mars turning from the fish:
Poison hidden under the heads of salmon,
Their war-chief hung with cord.

49
The advisers of the first monopoly,
The conquerors seduced for Malta:
Rhodes, Byzantium for them exposing their pole:
Land will fail the pursuers in flight.

50
When those of Hainault, of Ghent and of Brussels
Will see the siege laid before Langres:
Behind their flanks there will be cruel wars,
The ancient wound will do worse than enemies.

52
The blood of the just will commit a fault at London,
Burnt through lightning of twenty threes the six:
The ancient lady will fall from her high place,
Several of the same sect will be killed.

For several nights the earth will tremble:
In the spring two efforts in succession:
Corinth, Ephesus will swim in the two seas:
War stirred up by two valiant in combat.

53

The great plague of the maritime city
Will not cease until there be avenged the death
Of the just blood, condemned for a price without crime,
Of the great lady outraged by pretense.

54.

Because of people strange, and distant from the Romans
Their great city much troubled after water:
Daughter handless, domain too different,
Chief taken, lock not having been picked.

55

In the conflict the great one who was worth little
At his end will perform a marvelous deed:
While Adria will see what he was lacking,
During the banquet the proud one stabbed.

56

One whom neither plague nor steel knew how to finish,
Death on the summit of the hills struck from the sky:
The abbot will die when he will see ruined
Those of the wreck wishing to seize the rock.

57

Before the conflict the great wall will fall,
The great one to death, death too sudden and lamented,
Born imperfect: the greater part will swim:
Near the river the land stained with blood.

58

With neither foot nor hand because of sharp and strong tooth
Through the crowd to the fort of the pork and the elder born:
Near the portal treacherous proceeds,
Moon shining, little great one led off.

59

Gallic fleet through support of the great guard
Of the great Neptune, and his trident soldiers,
Provence reddened to sustain a great band:
More at Narbonne, because of javelins and darts.

60

The Punic faith broken in the East,
Ganges, Jordan, and Rhone, Loire, and Tagus will change:
When the hunger of the mule will be satiated,
Fleet sprinkles, blood and bodies will swim.

61

Bravo, ye of Tamins, Gironde and La Rochelle:
O Trojan blood! Mars at the port of the arrow
Behind the river the ladder put to the fort,
Points to fire great murder on the breach.

62

Mabus then will soon die, there will come
Of people and beasts a horrible rout:
Then suddenly one will see vengeance,
Hundred, hand, thirst, hunger when the comet will run.

64

The Gauls Ausonia will subjugate very little,
Po, Marne and Seine Parma will make drunk:
He who will prepare the great wall against them,
He will lose his life from the least at the wall.

64

The people of Geneva drying up with hunger, with thirst,
Hope at hand will come to fail:
On the point of trembling will be the law of him of the Cevennes,
Fleet at the great port cannot be received.

65

The sloping park great calamity
To be done through Hesperia and Insubria:
The fire in the ship, plague and captivity,
Mercury in Sagittarius Saturn will fade.

66

Through great dangers the captive escaped:
In a short time great his fortune changed.
In the palace the people are trapped,
Through good omen the city besieged.

67

The blond one will come to compromise the fork-nosed one
Through the duel and will chase him out:

The exiles within he will have restored,
Committing the strongest to the marine places.

68

The efforts of Aquilon will be great:
The gate on the Ocean will be opened,
The kingdom on the Isle will be restored:
London will tremble discovered by sail.

69

The Gallic King through his Celtic right arm
Seeing the discord of the great Monarchy:
He will cause his scepter to flourish over the three parts,
Against the cope of the great Hierarchy.

70

The dart from the sky will make its extension,
Deaths speaking: great execution.
The stone in the tree, the proud nation restored,
Noise, human monster, purge expiation.

71

The exiles will come into Sicily
To deliver form hunger the strange nation:
At daybreak the Celts will fail them:
Life remains by reason: the King joins.

72

Celtic army vexed in Italy
On all sides conflict and great loss:
Romans fled, O Gaul repelled!
Near the Ticino, Rubicon uncertain battle.

73

The shore of Lake Garda to Lake Fucino,
Taken from the Lake of Geneva to the port of L'Orguion:
Born with three arms the predicted warlike image,
Through three crowns to the great Endymion.

74

From Sens, from Autun they will come as far as the Rhone
To pass beyond towards the Pyrenees mountains:
The nation to leave the March of Ancona:
By land and sea it will be followed by great suites.

75

On the pipe of the air-vent floor:
So high will the bushel of wheat rise,
That man will be eating his fellow man.

76.
Lightning in Burgundy will perform a portentous deed,
One which could never have been done by skill,
Sexton made lame by their senate
Will make the affair known to the enemies.

77
Hurled back through bows, fires, pitch and by fires:
Cries, howls heard at midnight:
Within they are place on the broken ramparts,
The traitors fled by the underground passages.

78.
The great Neptune of the deep of the sea
With Punic race and Gallic blood mixed.
The Isles bled, because of the tardy rowing:
More harm will it do him than the ill-concealed secret.

79
The beard frizzled and black through skill
Will subjugate the cruel and proud people:
The great Chyren will remove from far away
All those captured by the banner of Selin

80
After the conflict by the eloquence of the wounded one
For a short time a soft rest is contrived:
The great ones are not to be allowed deliverance at all:
They are restored by the enemies at the proper time.

81
Through fire from the sky the city almost burned:
The Urn threatens Deucalion again:
Sardinia vexed by the Punic foist,
After Libra will leave her Phaethon.

82
Through hunger the prey will make the wolf prisoner,
The aggressor then in extreme distress.
The heir having the last one before him,
The great one does not escape in the middle of the crowd.

83

The large trade of a great Lyons changed,
The greater part turns to pristine ruin
Prey to the soldiers swept away by pillage:
Through the Jura mountain and Suevia drizzle.

84

Between Campania, Siena, Florence, Tuscany,
Six months nine days without a drop of rain:
The strange tongue in the Dalmatian land,
It will overrun, devastating the entire land.

85

The old full beard under the severe statute
Made at Lyon over the Celtic Eagle:
The little great one perseveres too far:
Noise of arms in the sky: Ligurian sea red.

86

Wreck for the fleet near the Adriatic Sea:
The land trembles stirred up upon the air placed on land:
Egypt trembles Mahometan increase,
The Herald surrendering himself is appointed to cry out.

87

After there will come from the outermost countries
A German Prince, upon the golden throne:
The servitude and waters met,
The lady serves, her time no longer adored.

88

The circuit of the great ruinous deed,
The seventh name of the fifth will be:
Of a third greater the stranger warlike:
Sheep, Paris, Aix will not guarantee.

89

One day the two great masters will be friends,
Their great power will be seen increased:
The new land will be at its high peak,
To the bloody one the number recounted.

90

Though life and death the realm of Hungary changed:
The law will be more harsh than service:
Their great city cries out with howls and laments,

Castor and Pollux enemies in the arena.

91.
At sunrise one will see a great fire,
Noise and light extending towards Aquilon:
Within the circle death and one will hear cries,
Through steel, fire, famine, death awaiting them.

92
Fire color of gold from the sky seen on earth:
Heir struck from on high, marvelous deed done:
Great human murder: the nephew of the great one taken,
Deaths spectacular the proud one escaped.

93
Very near the Tiber presses Death:
Shortly before great inundation:
The chief of the ship taken, thrown into the bilge:
Castle, palace in conflagration.

94
Great Po, great evil will be received through Gauls,
Vain terror to the maritime Lion:
People will pass by the sea in infinite numbers,
Without a quarter of a million escaping.

95
The populous places will be uninhabitable:
Great discord to obtain fields:
Realms delivered to prudent incapable ones:
Then for the great brothers dissension and death.

96
Burning torch will be seen in the sky at night
Near the end and beginning of the Rhone:
Famine, steel: the relief provided late,
Persia turns to invade Macedonia.

97
Roman Pontiff beware of approaching
The city that two rivers flow through,
Near there your blood will come to spurt,
You and yours when the rose will flourish.

98
The one whose face is splattered with the blood

Of the victim nearly sacrificed:
Jupiter in Leon, omen through presage:
To be put to death then for the bride.

99
Roman land as the omen interpreted
Will be vexed too much by the Gallic people:
But the Celtic nation will fear the hour,
The fleet has been pushed too far by the north wind.

100
Within the isles a very horrible uproar,
One will hear only a party of war,
So great will be the insult of the plunderers
That they will come to be joined in the great league.

Century III

This is the third Century of Nostradamus.

[Note: This is not my translation. It was found on the Internet a few years ago on a site that no longer exists. I have cleaned up gross misspellings and typos but have otherwise left the text intact. At some point, I hope to update with what I consider better translation. Due to lack of time on my part, this will have to suffice for now.]

For the French version, go to Nostradamiana at:
http://www.astrologer.ru:8003/Nostradamiana/centuries-eng.html

---

CENTURY III

1
After combat and naval battle,
The great Neptune in his highest belfry:
Red adversary will become pale with fear,
Putting the great Ocean in dread.

2
The divine word will give to the sustenance,
Including heaven, earth, gold hidden in the mystic milk:
Body, soul, spirit having all power,
As much under its feet as the Heavenly see.

3

Mars and Mercury, and the silver joined together,
Towards the south extreme drought:
In the depths of Asia one will say the earth trembles,
Corinth, Ephesus then in perplexity.

4
When they will be close the lunar ones will fail,
From one another not greatly distant,
Cold, dryness, danger towards the frontiers,
Even where the oracle has had its beginning.

5
Near, far the failure of the two great luminaries
Which will occur between April and March.
Oh, what a loss! but two great good-natured ones
By land and sea will relieve all parts.

6
Within the closed temple the lightning will enter,
The citizens within their fort injured:
Horses, cattle, men, the wave will touch the wall,
Through famine, drought, under the weakest armed.

7
The fugitives, fire from the sky on the pikes:
Conflict near the ravens frolicking,
From land they cry for aid and heavenly relief,
When the combatants will be near the walls.

8
The Cimbri joined with their neighbors
Will come to ravage almost Spain:
Peoples gathered in Guienne and Limousin
Will be in league, and will bear them company.

9
Bordeaux, Rouen and La Rochelle joined
Will hold around the great Ocean sea,
English, Bretons and the Flemings allied
Will chase them as far as Roanne.

10
Greater calamity of blood and famine,
Seven times it approaches the marine shore:
Monaco from hunger, place captured, captivity,
The great one led crunching in a metaled cage.

11
The arms to fight in the sky a long time,
The tree in the middle of the city fallen:
Sacred bough clipped, steel, in the face of the firebrand,
Then the monarch of Adria fallen.

12
Because of the swelling of the Ebro, Po, Tagus, Tiber and Rhône
And because of the pond of Geneva and Arezzo,
The two great chiefs and cities of the Garonne,
Taken, dead, drowned: human booty divided.

13.
Through lightning in the arch gold and silver melted,
Of two captives one will eat the other:
The greatest one of the city stretched out,
When submerged the fleet will swim.

14
Through the branch of the valiant personage
Of lowest France: because of the unhappy father
Honors, riches, travail in his old age,
For having believed the advice of a simple man.

15
The realm, will change in heart, vigor and glory,
In all points having its adversary opposed:
Then through death France an infancy will subjugate,
A great Regent will then be more contrary.

16
An English prince Marc in his heavenly heart
Will want to pursue his prosperous fortune,
Of the two duels one will pierce his gall:
Hated by him well loved by his mother.

17
Mount Aventine will be seen to burn at night:
The sky very suddenly dark in Flanders:
When the monarch will chase his nephew,
Then Church people will commit scandals.

18
After the rather long rain milk,

In several places in Reims the sky touched:
Alas, what a bloody murder is prepared near them,
Fathers and sons Kings will not dare approach.

19
In Lucca it will come to rain blood and milk,
Shortly before a change of praetor:
Great plague and war, famine and drought will be made visible
Far away where their prince and rector will die.

20
Through the regions of the great river Guadalquivir
Deep in Iberia to the Kingdom of Grenada
Crosses beaten back by the Mahometan peoples
One of Cordova will betray his country

21
In the Conca by the Adriatic Sea
There will appear a horrible fish,
With face human and its end aquatic,
Which will be taken without the hook.

22
Six days the attack made before the city:
Battle will be given strong and harsh:
Three will surrender it, and to them pardon:
The rest to fire and to bloody slicing and cutting.

23
If, France, you pass beyond the Ligurian Sea,
You will see yourself shut up in islands and seas:
Mahomet contrary, more so the Adriatic Sea:
You will gnaw the bones of horses and asses.

24
Great confusion in the enterprise,
Loss of people, countless treasure:
You ought not to extend further there.
France, let what I say be remembered.

25
He who will attain to the kingdom of Navarre
When Sicily and Naples will be joined:
He will hold Bigorre and Landes through Foix and Oloron
From one who will be too closely allied with Spain.

26
They will prepare idols of Kings and Princes,
Soothsayers and empty prophets elevated:
Horn, victim of gold, and azure, dazzling,
The soothsayers will be interpreted.

27
Libyan Prince powerful in the West
Will come to inflame very much French with Arabian.
Learned in letters condescending he will
Translate the Arabian language into French.

28
Of land weak and parentage poor,
Through piece and peace he will attain to the empire.
For a long time a young female to reign,
Never has one so bad come upon the kingdom.

29
The two nephews brought up in diverse places:
Naval battle, land, fathers fallen:
They will come to be elevated very high in making war
To avenge the injury, enemies succumbed.

30
He who during the struggle with steel in the deed of war
Will have carried off the prize from on greater than he:
By night six will carry the grudge to his bed,
Without armor he will surprised suddenly.

31
On the field of Media, of Arabia and of Armenia
Two great armies will assemble thrice:
The host near the bank of the Araxes,
They will fall in the land of the great Suleiman.

32
The great tomb of the people of Aquitaine
Will approach near to Tuscany,
When Mars will be in the corner of Germany
And in the land of the Mantuan people.

33
In the city where the wolf will enter,
Very near there will the enemies be:
Foreign army will spoil a great country.

The friends will pass at the wall and Alps.

34
When the eclipse of the Sun will then be,
The monster will be seen in full day:
Quite otherwise will one interpret it,
High price unguarded: none will have foreseen it.

35
From the very depths of the West of Europe,
A young child will be born of poor people,
He who by his tongue will seduce a great troop:
His fame will increase towards the realm of the East.

36
Buried apoplectic not dead,
He will be found to have his hands eaten:
When the city will condemn the heretic,
He who it seemed to them had changed their laws.

37
The speech delivered before the attack,
Milan taken by the Eagle through deceptive ambushes:
Ancient wall driven in by cannons,
Through fire and blood few given quarter.

38
The Gallic people and a foreign nation
Beyond the mountains, dead, captured and killed:
In the contrary month and near vintage time,
Through the Lords drawn up in accord.

39
The seven in three months in agreement
To subjugate the Apennine Alps:
But the tempest and cowardly Ligurian,
Destroys them in sudden ruins.

40
The great theater will come to be set up again:
The dice cast and the snares already laid.
Too much the first one will come to tire in the death knell,
Prostrated by arches already a long time split.

41
Hunchback will be elected by the council,

A more hideous monster not seen on earth,
The willing blow will put out his eye:
The traitor to the King received as faithful.

42
The child will be born with two teeth in his mouth,
Stones will fall during the rain in Tuscany:
A few years after there will be neither wheat nor barley,
To satiate those who will faint from hunger.

43
People from around the Tarn, Lot and Garonne
Beware of passing the Apennine mountains:
Your tomb near Rome and Ancona,
The black frizzled beard will have a trophy set up.

44
When the animal domesticated by man
After great pains and leaps will come to speak:
The lightning to the virgin will be very harmful,
Taken from earth and suspended in the air.

45
The five strangers entered in the temple,
Their blood will come to pollute the land:
To the Toulousans it will be a very hard example
Of one who will come to exterminate their laws.

46
The sky (of Plancus' city) forebodes to us
Through clear signs and fixed stars,
That the time of its sudden change is approaching,
Neither for its good, nor for its evils.

47
The old monarch chased out of his realm
Will go to the East asking for its help:
For fear of the crosses he will fold his banner:
To Mitylene he will go through port and by land.

48
Seven hundred captives bound roughly.
Lots drawn for the half to be murdered:
The hope at hand will come very promptly
But not as soon as the fifteenth death.

49

Gallic realm, you will be much changed:
To a foreign place is the empire transferred:
You will be set up amidst other customs and laws:
Rouen and Chartres will do much of the worst to you.

50

The republic of the great city
Will not want to consent to the great severity:
King summoned by trumpet to go out,
The ladder at the wall, the city will repent.

51

Paris conspires to commit a great murder
Blois will cause it to be fully carried out:
Those of Orléans will want to replace their chief,
Angers, Troyes, Langres will commit a misdeed against them.

52

In Campania there will be a very long rain,
In Apulia very great drought.
The Cock will see the Eagle, its wing poorly finished,
By the Lion will it be put into extremity.

53

When the greatest one will carry off the prize
Of Nuremberg, of Augsburg, and those of Bâle
Through Cologne the chief Frankfort retaken
They will cross through Flanders right into Gaul.

54

One of the greatest ones will flee to Spain
Which will thereafter come to bleed in a long wound:
Armies passing over the high mountains,
Devastating all, and then to reign in peace.

55

In the year that one eye will reign in France,
The court will be in very unpleasant trouble:
The great one of Blois will kill his friend:
The realm placed in harm and double doubt.

56

Montauban, Nîmes, Avignon and Béziers,
Plague, thunder and hail in the wake of Mars:
Of Paris bridge, Lyons wall, Montpellier,

After six hundreds and seven score three pairs.

57
Seven times will you see the British nation change,
Steeped in blood in 290 years:
Free not at all its support Germanic.
Aries doubt his Bastarnian pole.

58
Near the Rhine from the Noric mountains
Will be born a great one of people come too late,
One who will defend Sarmatia and the Pannonians,
One will not know what will have become of him.

59
Barbarian empire usurped by the third,
The greater part of his blood he will put to death:
Through senile death the fourth struck by him,
For fear that the blood through the blood be not dead.

60
Throughout all Asia (Minor) great proscription,
Even in Mysia, Lycia and Pamphilia.
Blood will be shed because of the absolution
Of a young black one filled with felony.

61
The great band and sect of crusaders
Will be arrayed in Mesopotamia:
Light company of the nearby river,
That such law will hold for an enemy.

62
Near the Douro by the closed Tyrian sea,
He will come to pierce the great Pyrenees mountains.
One hand shorter his opening glosses,
He will lead his traces to Carcassone.

63
The Roman power will be thoroughly abased,
Following in the footsteps of its great neighbor:
Hidden civil hatreds and debates
Will delay their follies for the buffoons.

64
The chief of Persia will occupy great Olchades,

The trireme fleet against the Mahometan people
From Parthia, and Media: and the Cyclades pillaged:
Long rest at the great Ionian port.

65
When the sepulcher of the great Roman is found,
The day after a Pontiff will be elected:
Scarcely will he be approved by the Senate
Poisoned, his blood in the sacred chalice.

66
The great Bailiff of Orléans put to death
Will be by one of blood revengeful:
Of death deserved he will not die, nor by chance:
He made captive poorly by his feet and hands.

67
A new sect of Philosophers
Despising death, gold, honors and riches
Will not be bordering upon the German mountains:
To follow them they will have power and crowds.

68
Leaderless people of Spain and Italy
Dead, overcome within the Peninsula:
Their dictator betrayed by irresponsible folly,
Swimming in blood everywhere in the latitude.

69
The great army led by a young man,
It will come to surrender itself into the hands of the enemies:
But the old one born to the half-pig,
He will cause Châlon and Mâcon to be friends.

70
The great Britain including England
Will come to be flooded very high by waters
The new League of Ausonia will make war,
So that they will come to strive against them.

71
Those in the isles long besieged
Will take vigor and force against their enemies:
Those outside dead overcome by hunger,
They will be put in greater hunger than ever before.

72
The good old man buried quite alive,
Near the great river through false suspicion:
The new old man ennobled by riches,
Captured on the road all his gold for ransom.

73
When the cripple will attain to the realm,
For his competitor he will have a near bastard:
He and the realm will become so very mangy
That before he recovers, it will be too late.

74
Naples, Florence, Faenza and Imola,
They will be on terms of such disagreement
As to delight in the wretches of Nola
Complaining of having mocked its chief.

75
Pau, Verona, Vicenza, Saragossa,
From distant swords lands wet with blood:
Very great plague will come with the great shell,
Relief near, and the remedies very far.

76
In Germany will be born diverse sects,
Coming very near happy paganism,
The heart captive and returns small,
They will return to paying the true tithe.

77
The third climate included under Aries
The year 1727 in October,
The King of Persia captured by those of Egypt:
Conflict, death, loss: to the cross great shame.

78
Captive of the Eastern seamen:
They will pass Gibraltar and Spain,
Present in Persia for the fearful new King.

79
The fatal everlasting order through the chain
Will come to turn through consistent order:
The chain of Marseilles will be broken:
The city taken, the enemy at the same time.

80

The worthy one chased out of the English realm,
The adviser through anger put to the fire:
His adherents will go so low to efface themselves
That the bastard will be half received.

81

The great shameless, audacious bawler,
He will be elected governor of the army:
The boldness of his contention,
The bridge broken, the city faint from fear.

82

Fréjus, Antibes, towns around Nice,
They will be thoroughly devastated by sea and by land:
The locusts by land and by sea the wind propitious,
Captured, dead, bound, pillaged without law of war.

83

The long hairs of Celtic Gaul
Accompanied by foreign nations,
They will make captive the people of Aquitaine,
For succumbing to their designs.

84

The great city will be thoroughly desolated,
Of the inhabitants not a single one will remain there:
Wall, sex, temple and virgin violated,
Through sword, fire, plague, cannon people will die.

85

The city taken through deceit and guile,
Taken in by means of a handsome youth:
Assault given by the Robine near the Aude,
He and all dead for having thoroughly deceived.

86

A chief of Ausonia will go to Spain
By sea, he will make a stop in Marseilles:
Before his death he will linger a long time:
After his death one will see a great marvel.

87

Gallic fleet, do not approach Corsica,
Less Sardinia, you will rue it:

Every one of you will die frustrated of the help of the cape:
You will swim in blood, captive you will not believe me.

88

From Barcelona a very great army by sea,
All Marseilles will tremble with terror:
Isles seized help shut off by sea,
Your traitor will swim on land.

89

At that time Cyprus will be frustrated
Of its relief by those of the Aegean Sea:
Old ones slaughtered: but by speeches and supplications
Their King seduced, Queen outraged more.

90

The great Satyr and Tiger of Hyrcania,
Gift presented to those of the Ocean:
A fleet's chief will set out from Carmania,
One who will take land at the Tyrren Phocaean.

91

The tree which had long been dead and withered,
In one night it will come to grow green again:
The Cronian King sick, Prince with club foot,
Feared by his enemies he will make his sail bound.

92

The world near the last period,
Saturn will come back again late:
Empire transferred towards the Dusky nation,
The eye plucked out by the Goshawk at Narbonne.

93

In Avignon the chief of the whole empire
Will make a stop on the way to desolated Paris:
Tricast will hold the anger of Hannibal:
Lyons will be poorly consoled for the change.

94

For five hundred years more one will keep count of him
Who was the ornament of his time:
Then suddenly great light will he give,
He who for this century will render them very satisfied.

95

The law of More will be seen to decline:
After another much more seductive:
Dnieper first will come to give way:
Through gifts and tongue another more attractive.

96
The Chief of Fossano will have his throat cut
By the leader of the bloodhound and greyhound:
The deed executed by those of the Tarpeian Rock,
Saturn in Leo February 13.

97
New law to occupy the new land
Towards Syria, Judea and Palestine:
The great barbarian empire to decay,
Before the Moon completes it cycle.

98
Two royal brothers will wage war so fiercely
That between them the war will be so mortal
That both will occupy the strong places:
Their great quarrel will fill realm and life.

99
In the grassy fields of Alleins and Vernègues
Of the Lubéron range near the Durance,
The conflict will be very sharp for both armies,
Mesopotamia will fail in France.

100
The last one honored amongst the Gauls,
Over the enemy man will he be victorious:
Force and land in a moment explored,
When the envious one will die from an arrow shot.

Century IV

This is the fourth Century by Nostradamus. The first 53 quatrains were originally published in 1555. The complete century appeared for the first time in 1557.

[Note: This is not my translation. It was found on the Internet a few years ago on a site that no longer exists. I have cleaned up gross misspellings and typos but have otherwise left the text intact. At some point, I hope to update with what I consider better translation. Due to lack of time on my part, this will have to suffice for now.]

For the French version, go to Nostradamiana at:
http://www.astrologer.ru:8003/Nostradamiana/centuries-eng.html

---

CENTURY IV

1

That of the remainder of blood unshed:
Venice demands that relief be given:
After having waited a very long time,
City delivered up at the first sound of the horn.

2

Because of death France will take to making a journey,
Fleet by sea, marching over the Pyrenees Mountains,
Spain in trouble, military people marching:
Some of the greatest Ladies carried off to France.

3

From Arras and Bourges many banners of Dusky Ones,
A greater number of Gascons to fight on foot,
Those along the Rhône will bleed the Spanish:
Near the mountain where Sagunto sits.

4

The impotent Prince angry, complaints and quarrels,
Rape and pillage, by cocks and Africans:
Great it is by land, by sea infinite sails,
Italy alone will be chasing Celts.

5

Cross, peace, under one the divine word accomplished,
Spain and Gaul will be united together:
Great disaster near, and combat very bitter:
No heart will be so hardy as not to tremble.

6

By the new clothes after the find is made,
Malicious plot and machination:
First will die he who will prove it,
Color Venetian trap.

7

The minor son of the great and hated Prince,

He will have a great touch of leprosy at the age of twenty:
Of grief his mother will die very sad and emaciated,
And he will die where the loose flesh falls.

8
The great city by prompt and sudden assault
Surprised at night, guards interrupted:
The guards and watches of Saint-Quentin
Slaughtered, guards and the portals broken.

9
The chief of the army in the middle of the crowd
Will be wounded by an arrow shot in the thighs,
When Geneva in tears and distress
Will be betrayed by Lausanne and the Swiss.

10
The young Prince falsely accused
Will plunge the army into trouble and quarrels:
The chief murdered for his support,
Scepter to pacify: then to cure scrofula.

11.
He who will have the government of the great cope
Will be prevailed upon to perform several deeds:
The twelve red one who will come to soil the cloth,
Under murder, murder will come to be perpetrated.

12
The greater army put to flight in disorder,
Scarcely further will it be pursued:
Army reassembled and the legion reduced,
Then it will be chased out completely from the Gauls.

13
News of the greater loss reported,
The report will astonish the army:
Troops united against the revolted:
The double phalanx will abandon the great one.

14
The sudden death of the first personage
Will have caused a change and put another in the sovereignty:
Soon, late come so high and of low age,
Such by land and sea that it will be necessary to fear him.

15
From where they will think to make famine come,
From there will come the surfeit:
The eye of the sea through canine greed
For the one the other will give oil and wheat.

16
The city of liberty made servile:
Made the asylum of profligates and dreamers.
The King changed to them not so violent:
From one hundred become more than a thousand.

17
To change at Beaune, Nuits, Châlon and Dijon,
The duke wishing to improve the Carmelite [nun]
Marching near the river, fish, diver's beak
Will see the tail: the gate will be locked.

18
Some of those most lettered in the celestial facts
Will be condemned by illiterate princes:
Punished by Edict, hunted, like criminals,
And put to death wherever they will be found.

19
Before Rouen the siege laid by the Insubrians,
By land and sea the passages shut up:
By Hainaut and Flanders, by Ghent and those of Liége
Through cloaked gifts they will ravage the shores.

20
Peace and plenty for a long time the place will praise:
Throughout his realm the fleur-de-lis deserted:
Bodies dead by water, land one will bring there,
Vainly awaiting the good fortune to be buried there.

21
The change will be very difficult:
City and province will gain by the change:
Heart high, prudent established, chased out one cunning,
Sea, land, people will change their state.

22
The great army will be chased out,
In one moment it will be needed by the King:
The faith promised from afar will be broken,

He will be seen naked in pitiful disorder.

23
The legion in the marine fleet
Will burn lime, lodestone sulfur and pitch:
The long rest in the secure place:
Port Selyn and Monaco, fire will consume them.

24
Beneath the holy earth of a soul the faint voice heard,
Human flame seen to shine as divine:
It will cause the earth to be stained with the blood of the monks,
And to destroy the holy temples for the impure ones.

25
Lofty bodies endlessly visible to the eye,
Through these reasons they will come to obscure:
Body, forehead included, sense and head invisible,
Diminishing the sacred prayers.

26
The great swarm of bees will arise,
Such that one will not know whence they have come;
By night the ambush, the sentinel under the vines
City delivered by five babblers not naked.

27
Salon, Tarascon, Mausol, the arch of SEX.,
Where the pyramid is still standing:
They will come to deliver the Prince of Annemark,
Redemption reviled in the temple of Artemis.

28
When Venus will be covered by the Sun,
Under the splendor will be a hidden form.
Mercury will have exposed them to the fire,
Through warlike noise it will be insulted.

29
The Sun hidden eclipsed by Mercury
Will be placed only second in the sky:
Of Vulcan Hermes will be made into food,
The Sun will be seen pure, glowing red and golden.

30
Eleven more times the Moon the Sun will not want,

All raised and lowered by degree:
And put so low that one will stitch little gold:
Such that after famine plague, the secret uncovered.

31
The Moon in the full of night over the high mountain,
The new sage with a lone brain sees it:
By his disciples invited to be immortal,
Eyes to the south. Hands in bosoms, bodies in the fire.

32
In the places and times of flesh giving way to fish,
The communal law will be made in opposition:
It will hold strongly the old ones, then removed from the midst,
Loving of Everything in Common put far behind.

33
Jupiter joined more to Venus than to the Moon
Appearing with white fullness:
Venus hidden under the whiteness of Neptune
Struck by Mars through the white stew.

34
The great one of the foreign land led captive,
Chained in gold offered to King Chyren:
He who in Ausonia, Milan will lose the war,
And all his army put to fire and sword.

35
The fire put out the virgins will betray
The greater part of the new band:
Lightning in sword and lance the lone Kings will guard
Etruria and Corsica, by night throat cut.

36
The new sports set up again in Gaul,
After victory in the Insubrian campaign:
Mountains of Hesperia, the great ones tied and trussed up:
Romania and Spain to tremble with fear.

37
The Gaul will come to penetrate the mountains by leaps:
He will occupy the great place of Insubria:
His army to enter to the greatest depth,
Genoa and Monaco will drive back the red fleet.

38
While he will engross the Duke, King and Queen
With the captive Byzantine chief in Samothrace:
Before the assault one will eat the order:
Reverse side metaled will follow the trail of the blood.

39
The Rhodians will demand relief,
Through the neglect of its heirs abandoned.
The Arab empire will reveal its course,
The cause set right again by Hesperia.

40
The fortresses of the besieged shut up,
Through gunpowder sunk into the abyss:
The traitors will all be stowed away alive,
Never did such a pitiful schism happen to the sextons.

41
Female sex captive as a hostage
Will come by night to deceive the guards:
The chief of the army deceived by her language
Will abandon her to the people, it will be pitiful to see.

42
Geneva and Langres through those of Chartres and Dôle
And through Grenoble captive at Montélimar
Seyssel, Lausanne, through fraudulent deceit,
They will betray them for sixty marks of gold.

43
Arms will be heard clashing in the sky:
That very same year the divine ones enemies:
They will want unjustly to discuss the holy laws:
Through lightning and war the complacent one put to death.

44
Two large ones of Mende, of Rodez and Milhau
Cahors, Limoges, Castres bad week
By night the entry, from Bordeaux an insult
Through Périgord at the peal of the bell.

45
Through conflict a King will abandon his realm:
The greatest chief will fail in time of need:
Dead, ruined few will escape it,

All cut up, one will be a witness to it.

46
The fact well defended by excellence,
Guard yourself Tours from your near ruin:
London and Nantes will make a defense through Reims
Not passing further in the time of the drizzle.

47
The savage black one when he will have tried
His bloody hand at fire, sword and drawn bows:
All of his people will be terribly frightened,
Seeing the greatest ones hung by neck and feet.

48
The fertile, spacious Ausonian plain
Will produce so many gadflies and locusts,
The solar brightness will become clouded,
All devoured, great plague to come from them.

49
Before the people blood will be shed,
Only from the high heavens will it come far:
But for a long time of one nothing will be heard,
The spirit of a lone one will come to bear witness against it.

50
Libra will see the Hesperias govern,
Holding the monarchy of heaven and earth:
No one will see the forces of Asia perished,
Only seven hold the hierarchy in order.

51
A Duke eager to follow his enemy
Will enter within impeding the phalanx:
Hurried on foot they will come to pursue so closely
That the day will see a conflict near Ganges.

52
In the besieged city men and woman to the walls,
Enemies outside the chief ready to surrender:
The wind will be strongly against the troops,
They will be driven away through lime, dust and ashes.

53
The fugitives and exiles recalled:

Fathers and sons great garnishing of the deep wells:
The cruel father and his people choked:
His far worse son submerged in the well.

54
Of the name which no Gallic King ever had
Never was there so fearful a thunderbolt,
Italy, Spain and the English trembling,
Very attentive to a woman and foreigners.

55
When the crow on the tower made of brick
For seven hours will continue to scream:
Death foretold, the statue stained with blood,
Tyrant murdered, people praying to their Gods.

56
After the victory of the raving tongue,
The spirit tempered in tranquillity and repose:
Throughout the conflict the bloody victor makes orations,
Roasting the tongue and the flesh and the bones.

57
Ignorant envy upheld before the great King,
He will propose forbidding the writings:
His wife not his wife tempted by another,
Twice two more neither skill nor cries.

58
To swallow the burning Sun in the throat,
The Etruscan land washed by human blood:
The chief pail of water, to lead his son away,
Captive lady conducted into Turkish land.

59
Two beset in burning fervor:
By thirst for two full cups extinguished,
The fort filed, and an old dreamer,
To the Genevans he will show the track from Nira.

60
The seven children left in hostage,
The third will come to slaughter his child:
Because of his son two will be pierced by the point,
Genoa, Florence, he will come to confuse them.

61

The old one mocked and deprived of his place,
By the foreigner who will suborn him:
Hands of his son eaten before his face,
His brother to Chartres, Orléans Rouen will betray.

62

A colonel with ambition plots,
He will seize the greatest army,
Against his Prince false invention,
And he will be discovered under his arbor.

63

The Celtic army against the mountaineers,
Those who will be learned and able in bird-calling:
Peasants will soon work fresh presses,
All hurled on the sword's edge.

64

The transgressor in bourgeois garb,
He will come to try the King with his offense:
Fifteen soldiers for the most part bandits,
Last of life and chief of his fortune.

65

Towards the deserter of the great fortress,
After he will have abandoned his place,
His adversary will exhibit very great prowess,
The Emperor soon dead will be condemned.

66

Under the feigned color of seven shaven heads
Diverse spies will be scattered:
Wells and fountains sprinkled with poisons,
At the fort of Genoa devourers of men.

67

The year that Saturn and Mars are equal fiery,
The air very dry parched long meteor:
Through secret fires a great place blazing from burning heat,
Little rain, warm wind, wars, incursions.

68

The two greatest ones of Asia and of Africa,
From the Rhine and Lower Danube they will be said to have come,
Cries, tears at Malta and the Ligurian side.

69

The exiles will hold the great city,
The citizens dead, murdered and driven out:
Those of Aquileia will promise Parma
To show them the entry through the untracked places.

70

Quite contiguous to the great Pyrenees mountains,
One to direct a great army against the Eagle:
Veins opened, forces exterminated,
As far as Pau will he come to chase the chief.

71

In place of the bride the daughters slaughtered,
Murder with great error no survivor to be:
Within the well vestals inundated,
The bride extinguished by a drink of Aconite.

72

Those of Nîmes through Agen and Lectoure
At Saint-Félix will hold their parliament:
Those of Bazas will come at the unhappy hour
To seize Condom and Marsan promptly.

73

The great nephew by force will test
The treaty made by the pusillanimous heart:
The Duke will try Ferrara and Asti,
When the pantomime will take place in the evening.

74

Those of lake Geneva and of Mâcon:
All assembled against those of Aquitaine:
Many Germans many more Swiss,
They will be routed along with those of the Humane.

75

Ready to fight one will desert,
The chief adversary will obtain the victory:
The rear guard will make a defense,
The faltering ones dead in the white territory.

76

Will be vexed, holding as far as the Rhône:
The union of Gascons and Bigorre

To betray the temple, the priest giving his sermon.

77
Selin monarch Italy peaceful,
Realms united by the Christian King of the World:
Dying he will want to lie in Blois soil,
After having chased the pirates from the sea.

78
The great army of the civil struggle,
By night Parma to the foreign one discovered,
Seventy-nine murdered in the town,
The foreigners all put to the sword.

79
Blood Royal flee, Monheurt, Mas, Aiguillon,
The Landes will be filled by Bordelais,
Navarre, Bigorre points and spurs,
Deep in hunger to devour acorns of the cork oak.

80
Near the great river, great ditch, earth drawn out,
In fifteen parts will the water be divided:
The city taken, fire, blood, cries, sad conflict,
And the greatest part involving the coliseum.

81
Promptly will one build a bridge of boats,
To pass the army of the great Belgian Prince:
Poured forth inside and not far from Brussels,
Passed beyond, seven cut up by pike.

82
A throng approaches coming from Slavonia,
The old Destroyer the city will ruin:
He will see his Romania quite desolated,
Then he will not know how to put out the great flame.

83
Combat by night the valiant captain
Conquered will flee few people conquered:
His people stirred up, sedition not in vain,
His own son will hold him besieged.

84
A great one of Auxerre will die very miserable,

Driven out by those who had been under him:
Put in chains, behind a strong cable,
In the year that Mars, Venus and Sun are in conjunction in summer.

85
The white coal will be chased by the black one,
Made prisoner led to the dung cart,
Moor Camel on twisted feet,
Then the younger one will blind the hobby falcon.

86
The year that Saturn will be conjoined in Aquarius
With the Sun, the very powerful King
Will be received and anointed at Reims and Aix,
After conquests he will murder the innocent.

87
A King's son learned in many languages,
Different from his senior in the realm:
His handsome father understood by the greater son,
He will cause his principal adherent to perish.

88
Anthony by name great by the filthy fact
Of Lousiness wasted to his end:
One who will want to be desirous of lead,
Passing the port he will be immersed by the elected one.

89
Thirty of London will conspire secretly
Against their King, the enterprise on the bridge:
He and his satellites will have a distaste for death,
A fair King elected, native of Frisia.

90
The two armies will be unable to unite at the walls,
In that instant Milan and Pavia to tremble:
Hunger, thirst, doubt will come to plague them very strongly
They will not have a single morsel of meat, bread or victuals.

91
For the Gallic Duke compelled to fight in the duel,
The ship of Melilla will not approach Monaco,
Wrongly accused, perpetual prison,
His son will strive to reign before his death.

92
The head of the valiant captain cut off,
It will be thrown before his adversary:
His body hung on the sail-yard of the ship,
Confused it will flee by oars against the wind.

93
A serpent seen near the royal bed,
It will be by the lady at night the dogs will not bark:
Then to be born in France a Prince so royal,
Come from heaven all the Princes will see him.

94
Two great brothers will be chased out of Spain,
The elder conquered under the Pyrenees mountains:
The sea to redden, Rhône, bloody Lake Geneva from Germany,
Narbonne, Béziers contaminated by Agde.

95
The realm left to two they will hold it very briefly,
Three years and seven months passed by they will make war:
The two Vestals will rebel in opposition,
Victor the younger in the land of Brittany.

96
The elder sister of the British Isle
Will be born fifteen years before her brother,
Because of her promise procuring verification,
She will succeed to the kingdom of the balance.

97
The year that Mercury, Mars, Venus in retrogression,
The line of the great Monarch will not fail:
Elected by the Portuguese people near Cadiz,
One who will come to grow very old in peace and reign.

98
Those of Alba will pass into Rome,
By means of Langres the multitude muffled up,
Marquis and Duke will pardon no man,
Fire, blood, smallpox no water the crops to fail.

99
The valiant elder son of the King's daughter,
He will hurl back the Celts very far,
Such that he will cast thunderbolts, so many in such an array

Few and distant, then deep into the Hesperias.

100
From the celestial fire on the Royal edifice,
When the light of Mars will go out,
Seven months great war, people dead through evil
Rouen, Evreux the King will not fail.

Century V

This is the fifth Century. It appeared for the first time in 1557.

[Note: This is not my translation. It was found on the Internet a few years ago on a site that no longer exists. I have cleaned up gross misspellings and typos but have otherwise left the text intact. At some point, I hope to update with what I consider better translation. Due to lack of time on my part, this will have to suffice for now.]

For the French version, go to Nostradamiana at:
http://www.astrologer.ru:8003/Nostradamiana/centuries-eng.html

---

CENTURY V

1
Before the coming of Celtic ruin,
In the temple two will parley
Pike and dagger to the heart of one mounted on the steed,
They will bury the great one without making any noise.

2
Seven conspirators at the banquet will cause to flash
The iron out of the ship against the three:
One will have the two fleets brought to the great one,
When through the evil the latter shoots him in the forehead.

3
The successor to the Duchy will come,
Very far beyond the Tuscan Sea:
A Gallic branch will hold Florence,
The nautical Frog in its bosom be agreement.

4
The large mastiff expelled from the city
Will be vexed by the strange alliance,

After having chased the stag to the fields
The wolf and the Bear will defy each other.

5
Under the shadowy pretense of removing servitude,
He will himself usurp the people and city:
He will do worse because of the deceit of the young prostitute,
Delivered in the field reading the false poem.

6
The Augur putting his hand upon the head of the King
Will come to pray for the peace of Italy:
He will come to move the scepter to his left hand,
From King he will become pacific Emperor.

7
The bones of the Triumvir will be found,
Looking for a deep enigmatic treasure:
Those from thereabouts will not be at rest,
Digging for this thing of marble and metallic lead.

8
There will be unleashed live fire, hidden death,
Horrible and frightful within the globes,
By night the city reduced to dust by the fleet,
The city afire, the enemy amenable.

9
The great arch demolished down to its base,
By the chief captive his friend forestalled,
He will be born of the lady with hairy forehead and face,
Then through cunning the Duke overtaken by death.

10
A Celtic chief wounded in the conflict
Seeing death overtaking his men near a cellar:
Pressed by blood and wounds and enemies,
And relief by four unknown ones.

11
The sea will not be passed over safely by those of the Sun,
Those of Venus will hold all Africa:
Saturn will no longer occupy their realm,
And the Asiatic part will change.

12

To near the Lake of Geneva will it be conducted,
By the foreign maiden wishing to betray the city:
Before its murder at Augsburg the great suite,
And those of the Rhine will come to invade it.

13
With great fury the Roman Belgian King
Will want to vex the barbarian with his phalanx:
Fury gnashing, he will chase the African people
From the Pannonias to the pillars of Hercules.

14
Saturn and Mars in Leo Spain captive,
By the African chief trapped in the conflict,
Near Malta, Herod taken alive,
And the Roman scepter will be struck down by the Cock.

15
The great Pontiff taken captive while navigating,
The great one thereafter to fail the clergy in tumult:
Second one elected absent his estate declines,
His favorite bastard to death broken on the wheel.

16
The Sabaean tear no longer at its high price,
Turning human flesh into ashes through death,
At the isle of Pharos disturbed by the Crusaders,
When at Rhodes will appear a hard phantom.

17
By night the King passing near an Alley,
He of Cyprus and the principal guard:
The King mistaken, the hand flees the length of the Rhône,
The conspirators will set out to put him to death.

18
The unhappy abandoned one will die of grief,
His conqueress will celebrate the hecatomb:
Pristine law, free edict drawn up,
The wall and the Prince falls on the seventh day.

19
The great Royal one of gold, augmented by brass,
The agreement broken, war opened by a young man:
People afflicted because of a lamented chief,
The land will be covered with barbarian blood.

20
The great army will pass beyond the Alps,
Shortly before will be born a monster scoundrel:
Prodigious and sudden he will turn
The great Tuscan to his nearest place.

21
By the death of the Latin Monarch,
Those whom he will have assisted through his reign:
The fire will light up again the booty divided,
Public death for the bold ones who incurred it.

22
Before the great one has given up the ghost at Rome,
Great terror for the foreign army:
The ambush by squadrons near Parma,
Then the two red ones will celebrate together.

23
The two contented ones will be united together,
When for the most part they will be conjoined with Mars:
The great one of Africa trembles in terror,
Duumvirate disjoined by the fleet.

24
The realm and law raised under Venus,
Saturn will have dominion over Jupiter:
The law and realm raised by the Sun,
Through those of Saturn it will suffer the worst.

25
The Arab Prince Mars, Sun, Venus, Leo,
The rule of the Church will succumb by sea:
Towards Persia very nearly a million men,
The true serpent will invade Byzantium and Egypt.

26
The slavish people through luck in war
Will become elevated to a very high degree:
They will change their Prince, one born a provincial,
An army raised in the mountains to pass over the sea.

27
Through fire and arms not far from the Black Sea,
He will come from Persia to occupy Trebizond:

Pharos, Mytilene to tremble, the Sun joyful,
The Adriatic Sea covered with Arab blood.

28
His arm hung and leg bound,
Face pale, dagger hidden in his bosom,
Three who will be sworn in the fray
Against the great one of Genoa will the steel be unleashed.

29
Liberty will not be recovered,
A proud, villainous, wicked black one will occupy it,
When the matter of the bridge will be opened,
The republic of Venice vexed by the Danube.

30
All around the great city
Soldiers will be lodged throughout the fields and towns:
To give the assault Paris, Rome incited,
Then upon the bridge great pillage will be carried out.

31
Through the Attic land fountain of wisdom,
At present the rose of the world:
The bridge ruined, and its great pre-eminence
Will be subjected, a wreck amidst the waves.

32
Where all is good, the Sun all beneficial and the Moon
Is abundant, its ruin approaches:
From the sky it advances to change your fortune.
In the same state as the seventh rock.

33
Of the principal ones of the city in rebelllon
Who will strive mightily to recover their liberty:
The males cut up, unhappy fray,
Cries, groans at Nantes pitiful to see.

34
From the deepest part of the English West
Where the head of the British isle is
A fleet will enter the Gironde through Blois,
Through wine and salt, fires hidden in the casks.

35

For the free city of the great Crescent sea,
Which still carries the stone in its stomach,
The English fleet will come under the drizzle
To seize a branch, war opened by the great one.

36
The sister's brother through the quarrel and deceit
Will come to mix dew in the mineral:
On the cake given to the slow old woman,
She dies tasting it she will be simple and rustic.

37
Three hundred will be in accord with one will
To come to the execution of their blow,
Twenty months after all memory
Their king betrayed simulating feigned hate.

38
He who will succeed the great monarch on his death
Will lead an illicit and wanton life:
Through nonchalance he will give way to all,
So that in the end the Salic law will fail.

39
Issued from the true branch of the fleur-de-lis,
Placed and lodged as heir of Etruria:
His ancient blood woven by long hand,
He will cause the escutcheon of Florence to bloom.

40
The blood royal will be so very mixed,
Gauls will be constrained by Hesperia:
One will wait until his term has expired,
And until the memory of his voice has perished.

41
Born in the shadows and during a dark day,
He will be sovereign in realm and goodness:
He will cause his blood to rise again in the ancient urn,
Renewing the age of gold for that of brass.

42
Mars raised to his highest belfry
Will cause the Savoyards to withdraw from France:
The Lombard people will cause very great terror
To those of the Eagle included under the Balance.

43
The great ruin of the holy things is not far off,
Provence, Naples, Sicily, Sées and Pons:
In Germany, at the Rhine and Cologne,
Vexed to death by all those of Mainz.

44
On sea the red one will be taken by pirates,
Because of him peace will be troubled:
Anger and greed will he expose through a false act,
The army doubled by the great Pontiff.

45
The great Empire will soon be desolated
And transferred to near the Ardennes:
The two bastards beheaded by the oldest one,
And Bronzebeard the hawk-nose will reign.

46
Quarrels and new schism by the red hats
When the Sabine will have been elected:
They will produce great sophism against him,
And Rome will be injured by those of Alba.

47
The great Arab will march far forward,
He will be betrayed by the Byzantians:
Ancient Rhodes will come to meet him,
And greater harm through the Austrian Hungarians.

48
After the great affliction of the scepter,
Two enemies will be defeated by them:
A fleet from Africa will appear before the Hungarians,
By land and sea horrible deeds will take place.

49
Not from Spain but from ancient France
Will one be elected for the trembling bark,
To the enemy will a promise be made,
He who will cause a cruel plague in his realm.

50
The year that the brothers of the lily come of age,
One of them will hold the great Romania:

The mountains to tremble, Latin passage opened,
Agreement to march against the fort of Armenia.

51
The people of Dacia, England, Poland
And of Bohemia will make a new league:
To pass beyond the pillars of Hercules,
The Barcelonians and Tuscans will prepare a cruel plot.

52
There will be a King who will give opposition,
The exiles raised over the realm:
The pure poor people to swim in blood,
And for a long time will he flourish under such a device.

53
The law of the Sun and of Venus in strife,
Appropriating the spirit of prophecy:
Neither the one nor the other will be understood,
The law of the great Messiah will hold through the Sun.

54
From beyond the Black Sea and great Tartary,
There will be a King who will come to see Gaul,
He will pierce through Alania and Armenia,
And within Byzantium will he leave his bloody rod.

55
In the country of Arabia Felix
There will be born one powerful in the law of Mahomet:
To vex Spain, to conquer Grenada,
And more by sea against the Ligurian people.

56
Through the death of the very old Pontiff
A Roman of good age will be elected,
Of him it will be said that he weakens his see,
But long will he sit and in biting activity.

57
There will go from Mont and Aventin,
One who through the hole will warn the army:
Between two rocks will the booty be taken,
Of Sectus' mausoleum the renown to fail.

58

By the aqueduct of Uzès over the Gard,
Through the forest and inaccessible mountain,
In the middle of the bridge there will be cut in the fist
The chief of Nîmes who will be very terrible.

59

Too long a stay for the English chief at Nîmes,
Towards Spain Redbeard to the rescue:
Many will die by war opened that day,
When a bearded star will fall in Artois.

60

By the shaven head a very bad choice will come to be made,
Overburdened he will not pass the gate:
He will speak with such great fury and rage,
That to fire and blood he will consign the entire sex.

61

The child of the great one not by his birth,
He will subjugate the high Apennine mountains:
He will cause all those of the balance to tremble,
And from the Pyrenees to Mont Cenis.

62

One will see blood to rain on the rocks,
Sun in the East, Saturn in the West:
Near Orgon war, at Rome great evil to be seen,
Ships sunk to the bottom, taken by Trident.

63

From the vain enterprise honor and undue complaint,
Boats tossed about among the Latins, cold, hunger, waves
Not far from the Tiber the land stained with blood,
And diverse plagues will be upon mankind.

64

Those assembled by the tranquillity of the great number,
By land and sea counsel countermanded:
Near Antonne Genoa, Nice in the shadow
Through fields and towns in revolt against the chief.

65

Come suddenly the terror will be great,
Hidden by the principal ones of the affair:
And the lady on the charcoal will no longer be in sight,
Thus little by little will the great ones be angered.

66

Under the ancient vestal edifices,
Not far from the ruined aqueduct:
The glittering metals are of the Sun and Moon,
The lamp of Trajan engraved with gold burning.

67

When the chief of Perugia will not venture his tunic
Sense under cover to strip himself quite naked:
Seven will be taken Aristocratic deed,
Father and son dead through a point in the collar.

68

In the Danube and of the Rhine will come to drink
The great Camel, not repenting it:
Those of the Rhône to tremble, and much more so those of the Loire,
and near the Alps the Cock will ruin him.

69

No longer will the great one be in his false sleep,
Uneasiness will come to replace tranquillity:
A phalanx of gold, azure and vermilion arrayed
To subjugate Africa and gnaw it to the bone,

70

Of the regions subject to the Balance,
They will trouble the mountains with great war,
Captives the entire sex enthralled and all Byzantium,
So that at dawn they will spread the news from land to land.

71

By the fury of one who will wait for the water,
By his great rage the entire army moved:
Seventeen boats loaded with the noble,
The messenger come late along the Rhône.

72

For the pleasure of the voluptuous edict,
One will mix poison in the faith:
Venus will be in a course so virtuous
As to becloud the whole quality of the Sun.

73

The Church of God will be persecuted,
And the holy Temples will be plundered,

The child will put his mother out in her shift,
Arabs will be allied with the Poles.

74
Of Trojan blood will be born a Germanic heart
Who will rise to very high power:
He will drive out the foreign Arabic people,
Returning the Church to its pristine pre-eminence.

75
He will rise high over the estate more to the right,
He will remain seated on the square stone,
Towards the south facing to his left,
The crooked staff in his hand his mouth sealed.

76
In a free place will he pitch his tent,
And he will not want to lodge in the cities:
Aix, Carpentras, L'Isle, Vaucluse Mont, Cavaillon,
Throughout all these places will he abolish his trace.

77
All degrees of Ecclesiastical honor
Will be changed to that of Jupiter and Quirinus:
The priest of Quirinus to one of Mars,
Then a King of France will make him one of Vulcan.

78
The two will not be united for very long,
And in thirteen years to the Barbarian Satrap:
On both sides they will cause such loss
That one will bless the Bark and its cope.

79
The sacred pomp will come to lower its wings,
Through the coming of the great legislator:
He will raise the humble, he will vex the rebels,
His like will not appear on this earth.

80
Ogmios will approach great Byzantium,
The Barbaric League will be driven out:
Of the two laws the heathen one will give way,
Barbarian and Frank in perpetual strife.

The royal bird over the city of the Sun,
Seven months in advance it will deliver a nocturnal omen:
The Eastern wall will fall lightning thunder,
Seven days the enemies directly to the gates.

82
At the conclusion of the treaty outside the fortress
Will not go he who is placed in despair:
When those of Arbois, of Langres against Bresse
Will have the mountains of Dôle an enemy ambush.

83
Those who will have undertaken to subvert,
An unparalleled realm, powerful and invincible:
They will act through deceit, nights three to warn,
When the greatest one will read his Bible at the table.

84
He will be born of the gulf and unmeasured city,
Born of obscure and dark family:
He who the revered power of the great King
Will want to destroy through Rouen and Evreux.

85
Through the Suevi and neighboring places,
They will be at war over the clouds:
Swarm of marine locusts and gnats,
The faults of Geneva will be laid quite bare.

86
Divided by the two heads and three arms,
The great city will be vexed by waters:
Some great ones among them led astray in exile,
Byzantium hard pressed by the head of Persia.

87
The year that Saturn is out of bondage,
In the Frank land he will be inundated by water:
Of Trojan blood will his marriage be,
And he will be confined safely be the Spaniards.

88
Through a frightful flood upon the sand,
A marine monster from other seas found:
Near the place will be made a refuge,
Holding Savona the slave of Turin.

89

Into Hungary through Bohemia, Navarre,
and under that banner holy insurrections:
By the fleur-de-lis legion carrying the bar,
Against Orléans they will cause disturbances.

90

In the Cyclades, in Perinthus and Larissa,
In Sparta and the entire Pelopennesus:
Very great famine, plague through false dust,
Nine months will it last and throughout the entire peninsula.

91

At the market that they call that of liars,
Of the entire Torrent and field of Athens:
They will be surprised by the light horses,
By those of Alba when Mars is in Leo and Saturn in Aquarius.

92

After the see has been held seventeen years,
Five will change within the same period of time:
Then one will be elected at the same time,
One who will not be too comfortable to the Romans.

93

Under the land of the round lunar globe,
When Mercury will be dominating:
The isle of Scotland will produce a luminary,
One who will put the English into confusion.

94

He will transfer into great Germany
Brabant and Flanders, Ghent, Bruges and Boulogne:
The truce feigned, the great Duke of Armenia
Will assail Vienna and Cologne.

95

The nautical oar will tempt the shadows,
Then it will come to stir up the great Empire:
In the Aegean Sea the impediments of wood
Obstructing the diverted Tyrrhenian Sea.

96

The rose upon the middle of the great world,
For new deeds public shedding of blood:

To speak the truth, one will have a closed mouth,
Then at the time of need the awaited one will come late.

97
The one born deformed suffocated in horror,
In the habitable city of the great King:
The severe edict of the captives revoked,
Hail and thunder, Condom inestimable.

98
At the forty-eighth climacteric degree,
At the end of Cancer very great dryness:
Fish in sea, river, lake boiled hectic,
Béarn, Bigorre in distress through fire from the sky.

99
Milan, Ferrara, Turin and Aquileia,
Capua, Brindisi vexed by the Celtic nation:
By the Lion and his Eagle's phalanx,
When the old British chief Rome will have.

100
The incendiary trapped in his own fire,
Of fire from the sky at Carcassonne and the Comminges:
Foix, Auch, Mazères, the high old man escaped,
Through those of Hesse and Thuringia, and some Saxons.

Century VI

This is the sixth Century by Nostradamus. It was first published in 1557.

[Note: This is not my translation. It was found on the Internet a few years ago on a site that no longer exists. I have cleaned up gross misspellings and typos but have otherwise left the text intact. At some point, I hope to update with what I consider better translation. Due to lack of time on my part, this will have to suffice for now.]

For the French version, go to Nostradamiana at:
http://www.astrologer.ru:8003/Nostradamiana/centuries-eng.html

----
CENTURY VI

1
Around the Pyrenees mountains a great throng
Of foreign people to aid the new King:

Near the great temple of Le Mas by the Garonne,
A Roman chief will fear him in the water.

2

In the year five hundred eighty more or less,
One will await a very strange century:
In the year seven hundred and three the heavens witness thereof,
That several kingdoms one to five will make a change.

3

The river that tries the new Celtic heir
Will be in great discord with the Empire:
The young Prince through the ecclesiastical people
Will remove the scepter of the crown of concord.

4

The Celtic river will change its course,
No longer will it include the city of Agrippina:
All changed except the old language,
Saturn, Leo, Mars, Cancer in plunder.

5

Very great famine through pestiferous wave,
Through long rain the length of the arctic pole:
Samarobryn one hundred leagues from the hemisphere,
The will live without law exempt from politics.

6

There will appear towards the North
Not far from Cancer the bearded star:
Susa, Siena, Boeotia, Eretria,
The great one of Rome will die, the night over.

7

Norway and Dacia and the British Isle
Will be vexed by the united brothers:
The Roman chief sprung from Gallic blood
And his forces hurled back into the forests.

8

Those who were in the realm for knowledge
Will become impoverished at the change of King:
Some exiled without support, having no gold,
The lettered and letters will not be at a high premium.

9

In the sacred temples scandals will be perpetrated,
They will be reckoned as honors and commendations:
Of one of whom they engrave medals of silver and of gold,
The end will be in very strange torments.

10

In a short time the temples with colors
Of white and black of the two intermixed:
Red and yellow ones will carry off theirs from them,
Blood, land, plague, famine, fire extinguished by water.

11

The seven branches will be reduced to three,
The elder ones will be surprised by death,
The two will be seduced to fratricide,
The conspirators will be dead while sleeping.

12

To raise forces to ascend to the empire
In the Vatican the Royal blood will hold fast:
Flemings, English, Spain with Aspire
Against Italy and France will he contend.

13

A doubtful one will not come far from the realm,
The greater part will want to uphold him:
A Capitol will not want him to reign at all,
He will be unable to bear his great burden.

14

Far from his land a King will lose the battle,
At once escaped, pursued, then captured,
Ignorant one taken under the golden mail,
Under false garb, and the enemy surprised.

15

Under the tomb will be found a Prince
Who will be valued above Nuremberg:
The Spanish King in Capricorn thin,
Deceived and betrayed by the great Wittenberg.

16

That which will be carried off by the young Hawk,
By the Normans of France and Picardy:
The black ones of the temple of the Black Forest place
Will make an inn and fire of Lombardy.

17
After the files the ass-drivers burned,
They will be obliged to change diverse garbs:
Those of Saturn burned by the millers,
Except the greater part which will not be covered.

18
The great King abandoned by the Physicians,
By fate not the Jew's art he remains alive,
He and his kindred pushed high in the realm,
Pardon given to the race which denies Christ.

19
The true flame will devour the lady
Who will want to put the Innocent Ones to the fire:
Before the assault the army is inflamed,
When in Seville a monster in beef will be seen.

20
The feigned union will be of short duration,
Some changed most reformed:
In the vessels people will be in suffering,
Then Rome will have a new Leopard.

21
When those of the arctic pole are united together,
Great terror and fear in the East:
Newly elected, the great trembling supported,
Rhodes, Byzantium stained with Barbarian blood.

22
Within the land of the great heavenly temple,
Nephew murdered at London through feigned peace:
The bark will then become schismatic,
Sham liberty will be proclaimed everywhere.

23
Coins depreciated by the spirit of the realm,
And people will be stirred up against their King:
New peace made, holy laws become worse,
Paris was never in so severe an array.

24
Mars and the scepter will be found conjoined
Under Cancer calamitous war:

Shortly afterwards a new King will be anointed,
One who for a long time will pacify the earth.

25
Through adverse Mars will the monarchy
Of the great fisherman be in ruinous trouble:
The young red black one will seize the hierarchy,
The traitors will act on a day of drizzle.

26
For four years the see will be held with some little good,
One libidinous in life will succeed to it:
Ravenna, Pisa and Verona will give support,
Longing to elevate the Papal cross.

27
Within the Isles of five rivers to one,
Through the expansion of the great Chyren Selin:
Through the drizzles in the air the fury of one,
Six escaped, hidden bundles of flax.

28
The great Celt will enter Rome,
Leading a throng of the exiled and banished:
The great Pastor will put to death every man
Who was united at the Alps for the cock.

29
The saintly widow hearing the news,
Of her offspring placed in perplexity and trouble:
He who will be instructed to appease the quarrels,
He will pile them up by his pursuit of the shaven heads.

30
Through the appearance of the feigned sanctity,
The siege will be betrayed to the enemies:
In the night when they trusted to sleep in safety,
Near Brabant will march those of Liège.

31
The King will find that which he desired so much
When the Prelate will be blamed unjustly:
His reply to the Duke will leave him dissatisfied,
He who in Milan will put several to death.

Beaten to death by rods for treason,
Captured he will be overcome through his disorder:
Frivolous counsel held out to the great captive,
When Berich will come to bite his nose in fury.

33
His last hand through sanguinary,
He will be unable to protect himself by sea:
Between two rivers he will fear the military hand,
The black and irate one will make him rue it.

34
The device of flying fire
Will come to trouble the great besieged chief:
Within there will be such sedition
That the profligate ones will be in despair.

35
Near the Bear and close to the white wool,
Aries, Taurus, Cancer, Leo, Virgo,
Mars, Jupiter, the Sun will burn a great plain,
Woods and cities letters hidden in the candle.

36
Neither good nor evil through terrestrial battle
Will reach the confines of Perugia,
Pisa to rebel, Florence to see an evil existence,
King by night wounded on a mule with black housing.

37
The ancient work will be finished,
Evil ruin will fall upon the great one from the roof:
Dead they will accuse an innocent one of the deed,
The guilty one hidden in the copse in the drizzle.

38
The enemies of peace to the profligates,
After having conquered Italy:
The bloodthirsty black one, red, will be exposed,
Fire, blood shed, water colored by blood.

39
The child of the realm through the capture of his father
Will be plundered to deliver him:
Near the Lake of Perugia the azure captive,
The hostage troop to become far too drunk.

40

To quench the great thirst the great one of Mainz
Will be deprived of his great dignity:
Those of Cologne will come to complain so loudly
That the great rump will be thrown into the Rhine.

41

The second chief of the realm of Annemark,
Through those of Frisia and of the British Isle,
Will spend more than one hundred thousand marks,
Exploiting in vain the voyage to Italy.

42

To Ogmios will be left the realm
Of the great Selin, who will in fact do more:
Throughout Italy will he extend his banner,
He will be ruled by a prudent deformed one.

43

For a long time will she remain uninhabited,
Around where the Seine and the Marne she comes to water:
Tried by the Thames and warriors,
The guards deceived in trusting in the repulse.

44

By night the Rainbow will appear for Nantes,
By marine arts they will stir up rain:
In the Gulf of Arabia a great fleet will plunge to the bottom,
In Saxony a monster will be born of a bear and a sow.

45

The very learned governor of the realm,
Not wishing to consent to the royal deed:
The fleet at Melilla through contrary wind
Will deliver him to his most disloyal one.

46

A just one will be sent back again into exile,
Through pestilence to the confines of Nonseggle,
His reply to the red one will cause him to be misled,
The King withdrawing to the Frog and the Eagle.

47

The two great ones assembled between two mountains
Will abandon their secret quarrel:

Brussels and Dôle overcome by Langres,
To execute their plague at Malines.

48
The too false and seductive sanctity,
Accompanied by an eloquent tongue:
The old city, and Parma too premature,
Florence and Siena they will render more desert.

49
The great Pontiff of the party of Mars
Will subjugate the confines of the Danube:
The cross to pursue, through sword hook or crook,
Captives, gold, jewels more than one hundred thousand rubies.

50
Within the pit will be found the bones,
Incest will be committed by the stepmother:
The state changed, they will demand fame and praise,
And they will have Mars attending as their star.

51
People assembled to see a new spectacle,
Princes and Kings amongst many bystanders,
Pillars walls to fall: but as by a miracle
The King saved and thirty of the ones present.

52
In place of the great one who will be condemned,
Outside the prison, his friend in his place:
The Trojan hope in six months joined, born dead,
The Sun in the urn rivers will be frozen.

53
The great Celtic Prelate suspected by the King,
By night in flight he will leave the realm:
Through a Duke fruitful for his great British King,
Byzantium to Cyprus and Tunis unsuspected.

54
At daybreak at the second crowing of the cock,
Those of Tunis, of Fez and of Bougie,
By the Arabs the King of Morocco captured,
The year sixteen hundred and seven, of the Liturgy.

55

By the appeased Duke in drawing up the contract,
Arabesque sail seen, sudden discovery:
Tripoli, Chios, and those of Trebizond,
Duke captured, the Black Sea and the city a desert.

56

The dreaded army of the Narbonne enemy
Will frighten very greatly the Hesperians:
Perpignan empty through the blind one of Arbon,
Then Barcelona by sea will take up the quarrel.

57

He who was well forward in the realm,
Having a red chief close to the hierarchy,
Harsh and cruel, and he will make himself much feared,
He will succeed to the sacred monarchy.

58

Between the two distant monarchs,
When the clear Sun is lost through Selin:
Great enmity between two indignant ones,
So that liberty is restored to the Isles and Siena.

59

The Lady in fury through rage of adultery,
She will come to conspire not to tell her Prince:
But soon will the blame be made known,
So that seventeen will be put to martyrdom.

60

The Prince outside his Celtic land
Will be betrayed, deceived by the interpreter:
Rouen, La Rochelle through those of Brittany
At the port of Blaye deceived by monk and priest.

61

The great carpet folded will not show
But by halved the greatest part of history:
Driven far out of the realm he will appear harsh,
So that everyone will come to believe in his warlike deed.

62

Too late both the flowers will be lost,
The serpent will not want to act against the law:
The forces of the Leaguers confounded by the French,
Savona, Albenga through Monaco great martyrdom.

63

The lady left alone in the realm
By the unique one extinguished first on the bed of honor:
Seven years will she be weeping in grief,
Then with great good fortune for the realm long life.

64

No peace agreed upon will be kept,
All the subscribers will act with deceit:
In peace and truce, land and sea in protest,
By Barcelona fleet seized with ingenuity.

65

Gray and brown in half-opened war,
By night they will be assaulted and pillaged:
The brown captured will pass through the lock,
His temple opened, two slipped in the plaster.

66

At the foundation of the new sect,
The bones of the great Roman will be found,
A sepulcher covered by marble will appear,
Earth to quake in April poorly buried.

67

Quite another one will attain to the great Empire,
Kindness distant more so happiness:
Ruled by one sprung not far from the brothel,
Realms to decay great bad luck.

68

When the soldiers in a seditious fury
Will cause steel to flash by night against their chief:
The enemy Alba acts with furious hand,
Then to vex Rome and seduce the principal ones.

69

The great pity will occur before long,
Those who gave will be obliged to take:
Naked, starving, withstanding cold and thirst,
To pass over the mountains committing a great scandal.

70

Chief of the world will the great Chyren be,
Plus Ultra behind, loved, feared, dreaded:

His fame and praise will go beyond the heavens,
And with the sole title of Victor will he be quite satisfied.

71
When they will come to give the last rites to the great King
Before he has entirely given up the ghost:
He who will come to grieve over him the least,
Through Lions, Eagles, cross crown sold.

72
Through feigned fury of divine emotion
The wife of the great one will be violated:
The judges wishing to condemn such a doctrine,
She is sacrificed a victim to the ignorant people.

73
In a great city a monk and artisan,
Lodged near the gate and walls,
Secret speaking emptily against Modena,
Betrayed for acting under the guise of nuptials.

74
She chased out will return to the realm,
Her enemies found to be conspirators:
More than ever her time will triumph,
Three and seventy to death very sure.

75
The great Pilot will be commissioned by the King,
To leave the fleet to fill a higher post:
Seven years after he will be in rebellion,
Venice will come to fear the Barbarian army.

76
The ancient city the creation of Antenor,
Being no longer able to bear the tyrant:
The feigned handle in the temple to cut a throat,
The people will come to put his followers to death.

77
Through the fraudulent victory of the deceived,
Two fleets one, German revolt:
The chief murdered and his son in the tent,
Florence and Imola pursued into Romania.

78

To proclaim the victory of the great expanding Selin:
By the Romans will the Eagle be demanded,
Pavia, Milan and Genoa will not consent thereto,
Then by themselves the great Lord claimed.

79
Near the Ticino the inhabitants of the Loire,
Garonne and Saône, the Seine, the Tain and Gironde:
They will erect a promontory beyond the mountains,
Conflict given, Po enlarged, submerged in the wave.

80
From Fez the realm will reach those of Europe,
Their city ablaze and the blade will cut:
The great one of Asia by land and sea with great troop,
So that blues and Pers[ians] the cross will pursue to death.

81
Tears, cries and laments, howls, terror,
Heart inhuman, cruel, black and chilly:
Lake of Geneva the Isles, of Genoa the notables,
Blood to pour out, wheat famine to none mercy.

82
Through the deserts of the free and wild place,
The nephew of the great Pontiff will come to wander:
Felled by seven with a heavy club,
By those who afterwards will occupy the Chalice.

83
He who will have so much honor and flattery
At his entry into Belgian Gaul:
A while after he will act very rudely,
And he will act very warlike against the flower.

84
The Lame One, he who lame could not reign in Sparta,
He will do much through seductive means:
So that by the short and long, he will be accused
Of making his perspective against the King.

85
The great city of Tarsus by the Gauls
Will be destroyed, all of the Turban captives:
Help by sea from the great one of Portugal,
First day of summer Urban's consecration.

86
The great Prelate one day after his dream,
Interpreted opposite to its meaning:
From Gascony a monk will come unexpectedly,
One who will cause the great prelate of Sens to be elected.

87
The election made in Frankfort
Will be voided, Milan will be opposed:
The follower closer will seem so very strong
That he will drive him out into the marshes beyond the Rhine.

88
A great realm will be left desolated,
Near the Ebro an assembly will be formed:
The Pyrenees mountains will console him,
When in May lands will be trembling.

89
Feet and hands bound between two boats,
Face anointed with honey, and sustained with milk:
Wasps and flies, paternal love vexed,
Cup-bearer to falsify, Chalice tried.

90
The stinking abominable disgrace,
After the deed he will be congratulated:
The great excuse for not being favorable,
That Neptune will not be persuaded to peace.

91
Of the leader of the naval war,
Red one unbridled, severe, horrible whim,
Captive escaped from the elder one in the bale,
When there will be born a son to the great Agrippa.

92
Prince of beauty so comely,
Around his head a plot, the second deed betrayed:
The city to the sword in dust the face burnt,
Through too great murder the head of the King hated.

93
The greedy prelate deceived by ambition,
He will come to reckon nothing too much for him:

He and his messengers completely trapped,
He who cut the wood sees all in reverse.

94

A King will be angry with the see-breakers,
When arms of war will be prohibited:
The poison tainted in the sugar for the strawberries,
Murdered by waters, dead, saying land, land.

95

Calumny against the cadet by the detractor,
When enormous and warlike deeds will take place:
The least part doubtful for the elder one,
And soon in the realm there will be partisan deeds.

96

Great city abandoned to the soldiers,
Never was mortal tumult so close to it:
Oh, what a hideous calamity draws near,
Except one offense nothing will be spared it.

97

At forty-five degrees the sky will burn,
Fire to approach the great new city:
In an instant a great scattered flame will leap up,
When one will want to demand proof of the Normans.

98

Ruin for the Volcae so very terrible with fear,
Their great city stained, pestilential deed:
To plunder Sun and Moon and to violate their temples:
And to redden the two rivers flowing with blood.

99

The learned enemy will find himself confused,
His great army sick, and defeated by ambushes,
The Pyrenees and Pennine Alps will be denied him,
Discovering near the river ancient jugs.

100

INCANTATION OF THE LAW AGAINST INEPT CRITICS
Let those who read this verse consider it profoundly,
Let the profane and the ignorant herd keep away:
And far away all Astrologers, Idiots and Barbarians,
May he who does otherwise be subject to the sacred rite.

Century VII

This is the seventh Century by Nostradamus. It contains only 42 quatrains. The first 40 were first published in 1557; the last two did not appear until 1568.

[Note: This is not my translation. It was found on the Internet a few years ago on a site that no longer exists. I have cleaned up gross misspellings and typos but have otherwise left the text intact. At some point, I hope to update with what I consider better translation. Due to lack of time on my part, this will have to suffice for now.]

For the French version, go to Nostradamiana at:
http://www.astrologer.ru:8003/Nostradamiana/centuries-eng.html

---

CENTURY VII

1
The arc of the treasure deceived by Achilles,
the quadrangle known to the procreators.
The invention will be known by the Royal deed;
a corpse seen hanging in the sight of the populace.

2
Opened by Mars Arles will not give war,
the soldiers will be astonished by night.
Black and white concealing indigo on land
under the false shadow you will see traitors sounded.

3
After the naval victory of France,
the people of Barcelona the Saillinons and those of Marseilles;
the robber of gold, the anvil enclosed in the ball,
the people of Ptolon will be party to the fraud.

4
The Duke of Langres besieged at Dôle
accompanied by people from Autun and Lyons.
Geneva, Augsburg allied to those of Mirandola,
to cross the mountains against the people of Ancona.

5
Some of the wine on the table will be spilt,
the third will not have that which he claimed.
Twice descended from the black one of Parma,

Perouse will do to Pisa that which he believed.

6
Naples, Palerma and all of Sicily
will be uninhabited through Barbarian hands.
Corsica, Salerno and the island of Sardinia,
hunger, plague, war the end of extended evils.

7
Upon the struggle of the great light horses,
it will be claimed that the great crescent is destroyed.
To kill by night, in the mountains,
dressed in shepherd's' clothing, red gulfs in the deep ditch.

8
Florense, flee, flee the nearest Roman,
at Fiesole will be conflict given:
blood shed, the greatest one take by the hand,
neither temple nor sex will be pardoned.

9
The lady in the absence of her great master
will be begged for love by the Viceroy.
Feigned promise and misfortune in love,
in the hands of the great Prince of Bar.

10
By the great Prince bordering Le Mans,
brave and valiant leader of the great army;
by land and sea with Bretons and Normans,
to pass Gibraltar and Barcelona to pillage the island.

11
eye, feet wounded rude disobedient;
strange and very bitter news to the lady;
more than five hundred of here people will be killed.

12
The great younger son will make an end of the war,
he assembles the pardoned before the gods;
Cahors and Moissac will go far from the prison,
a refusal at Lectoure, the people of Agen shaved.

13
From the marine tributary city,
the shaven head will take up the satrapy;

to chase the sordid man who will the be against him.
For fourteen years he will hold the tyranny.

14
He will come to expose the false topography,
the urns of the tombs will be opened.
Sect and holy philosophy to thrive,
black for white and the new for the old.

15
Before the city of the Insubrian lands,
for seven years the siege will be laid;
a very great king enters it,
the city is then free, away from its enemies.

16
The deep entry made by the great Queen
will make the place powerful and inaccessible;
the army of the three lions will be defeated
causing within a thing hideous and terrible.

17
The prince who has little pity of mercy
will come through death to change (and become) very knowledgeable.
The kingdom will be attended with great tranquillity,
when the great one will soon be fleeced.

18
The besieged will color their pacts,
but seven days later they will make a cruel exit:
thrown back inside, fire and blood, seven put to the ax
the lady who had woven the peace is a captive.

19
The fort at Nice will not engage in combat,
it will be overcome by shining metal.
This deed will be debated for a long time,
strange and fearful for the citizens.

20
Ambassadors of the Tuscan language
will cross the Alps and the sea in April and May.
The man of the calf will deliver an oration,
not coming to wipe out the French way of life.

21

By the pestilential enmity of Languedoc,
the tyrant dissimulated will be driven out.
The bargain will be made on the bridge at Sorgues
to put to death both him and his follower

22
The citizens of Mesopotamia
angry with their friends from Tarraconne;
games, rites, banquets, every person asleep,
the vicar at Rhône, the city taken and those of Ausonia.

23
The Royal scepter will be forced to take
that which his predecessors had pledged.
Because they do not understand about the ring
when they come to sack the palace.

24
He who was buried will come out of the tomb,
He will cause the fort of the bridge to be tied in chains:
Poisoned with the spawn of a pimp,
the great one from Lorraine by the Marquis du Pont.

25
Through long war all the army exhausted,
so that they do not find money for the soldiers;
instead of gold or silver, they will come to coin leather,
Gallic brass, and the crescent sign of the Moon.

26
Foists and galleys around seven ships,
a mortal war will be let loose.
The leader from Madrid will receive a wound from arrows,
two escaped and five brought to land.

27
At the wall of Vasto the great cavalry
are impeded by the baggage near Ferrara.
At Turin they will speedily commit such robbery
that in the fort they will ravish their hostage.

28
The captain will lead a great herd
on the mountain closest to the enemy.
Surrounded by fire he makes such a way,
all escape except for thirty put on the spit.

29

The great one of Alba will come to rebel,
he will betray his great forebears.
The great man of Guise will come to vanquish him,
led captive with a monument erected.

30

The sack approaches, fire and great bloodshed.
Po the great rivers, the enterprise for the clowns;
after a long wait from Genoa and Nice,
Fossano, Turin the capture at Savigliano.

31

From Languedoc and Guienne more than ten
thousand will want to cross the Alps again.
The great Savoyards march against Brindisi,
Aquino and Bresse will come to drive them back.

32

From the bank of Montereale will be born one
who bores and calculates becoming a tyrant.
To raise a force in the marches of Milan,
to drain Faenza and Florence of gold and men

33

The kingdom stripped of its forces by fraud,
the fleet blockaded, passages for the spy;
two false friends will come to rally
to awaken hatred for a long time dormant.

34

The French nation will be in great grief,
vain and lighthearted, they will believe rash things.
No bread, salt, wine nor water, venom nor ale,
the greater one captured, hunger, cold and want.

35

The great fish will come to complain and weep
for having chosen, deceived concerning his age:
he will hardly want to remain with them,
he will be deceived by those (speaking) his own tongue.

36

God, the heavens, all the divine words in the waves,
carried by seven red-shaven heads to Byzantium:

against the anointed three hundred from Trebizond,
will make two laws, first horror then trust.

37
Ten sent to put the captain of the ship to death,
are altered by one that there is open revolt in the fleet.
Confusion, the leader and another stab and bite each other
at Lerins and the Hyères, ships, prow into the darkness.

38
The elder royal one on a frisky horse
will spur so fiercely that it will bolt.
Mouth, mouthful, foot complaining in the embrace;
dragged, pulled, to die horribly.

39
The leader of the French army
will expect to lose the main phalanx.
Upon the pavement of oats and slate
the foreign nation will be undermined through Genoa.

40
Within casks anointed outside with oil and grease
twenty-one will be shut before the harbor,
at second watch; through death they will do great deeds;
to win the gates and be killed by the watch.

41
The bones of the feet and the hands locked up,
because of the noise the house is uninhabited for a long time.
Digging in dreams they will be unearthed,
the house healthy in inhabited without noise.

42
Two newly arrived have seized the poison,
to pour it in the kitchen of the great Prince
By the scullion both are caught in the act,
taken he who thought to trouble the elder with death.

Epistle to Henry II

Here is an English translation of the Epistle to Henry II that was published as a preface to the last three centuries in 1568.

[Note: This is not my translation. It was found on the Internet a few years ago on a site that no longer exists. I have cleaned up gross misspellings and typos but have otherwise left the text intact. At some point, I hope to update with what I consider better translation. Due to lack of time on my part, this will have to suffice for now.]

For the French version, go to Nostradamiana at:
http://www.astrologer.ru:8003/Nostradamiana/centuries-eng.html

---

EPISTLE TO HENRY II

TO THE MOST INVINCIBLE
MOST POWERFUL AND MOST CHRISTIAN
HENRY, KING OF FRANCE THE SECOND:
MICHEL NOSTRADAMUS,
HIS VERY HUMBLE AND VERY OBEDIENT SERVANT AND SUBJECT,
WISHES VICTORY AND HAPPINESS

Ever since my long-beclouded face first presented itself before the immeasurable deity of your Majesty, O Most Christian and Most Victorious King, I have remained perpetually dazzled by that sovereign sight. I have never ceased to honor and venerate properly that date when I presented myself before a Majesty so singular and so humane. I have searched for some occasion on which to manifest high heart and stout courage, and thereby obtain even greater recognition of Your Most Serene Majesty. But I saw how obviously impossible it was for me to declare myself.

While I was seized with this singular desire to be transported suddenly from my long-beclouded obscurity to the illuminating presence of the first monarch of the universe, I was also long in doubt as to whom I would dedicate these last three Centuries of my prophecies, making up the thou- sand. After having meditated for a long time on an act of such rash audacity, I have ventured to address Your Majesty. I have not been daunted like those mentioned by that most grave author Plutarch, in his Life of Lycurgus, who were so astounded at the expense of the offerings and gifts brought as sacrifices to the temples of the immortal gods of that age, that they did not dare to present anything at all. Seeing your royal splendor to be accompanied by such an incomparable humanity, I have paid my address to it and not as those Kings of Persia whom one could neither stand before nor approach.

It is to a most prudent and most wise Prince that I have dedicated my nocturnal and prophetic calculations, which are composed rather out of a natural instinct, accompanied by a poetic furor, than according to the strict rules of poetry. Most of them have been integrated with astronomical calculations corresponding to the years, months and weeks of the regions, countries and most of the towns and cities of all Europe, including Africa and part of Asia, where most of all these coming events are to transpire. They are composed in a natural manner.

Indeed, someone, who would do well to blow his nose, may reply that the rhythm is as easy as the sense is difficult. That, O Most Humane king, is because most of the prophetic quatrains are so ticklish that there is no making way through them, nor is there any interpreting of them.

Nevertheless, I wanted to leave a record in writing of the years, towns, cities and regions in which most of the events will come to pass, even those of the year 1585 and of the year 1606, reckoning from the present time, which is March 14, 1557, and going far beyond to the events which will take place at the beginning of the seventh millenary, when, so far as my pro- found astronomical calculations and other knowledge have been able to make out, the adversaries of Jesus Christ and his Church will begin to multiply greatly.

I have calculated and composed all during choice hours of well-disposed days, and as accurately as I could, all when Minerva was free and not unfavorable. I have made computations for events over almost as long a period to come as that which has already passed, and by these they will know in all regions what is to happen in the course of time, just as it is writ- ten, with nothing superfluous added, although some may say, There can be no truth entirely determined concerning the future.

It is quite true, Sire, that my natural instinct has been inherited from my forebears, who did not believe in predicting, and that this is natural instinct has been adjusted and integrated with long calculations. At the same time, I freed my soul, mind and heart of all care, solicitude and vexation. All of these prerequisites for presaging I achieved in part by means of the brazen tripod.

There are some who would attribute to me that which is not mine at all. The eternal God alone, who is the thorough searcher of humane hearts, pious, just and merciful, is the true judge, and it is to him I pray to defend me from the calumny of evil men. These evil ones, in their slanderous way, would likewise want to inquire how all your most ancient progenitors, the Kings of France, have cured the scrofula, how those of other nations have cured the bite of snakes, how those of yet other nations have had a certain instinct for the art of divination and still others which would be too long to recite here.

Notwithstanding those who cannot contain the malignity of the evil spirit, as time elapses after my death, my writings will have more weight than during my lifetime. Should I, however, have made any errors in my calculation of dates, or prove unable to please everybody, I beg that your more than Imperial Majesty will forgive me. I protest before God and his Saints that I do not propose to insert any writings in this present Epistle that will be contrary to the true Catholic faith, whilst consulting the astronomical calculations to the best of my ability.

Such is the extent of time past, subject to correction by the most learned judgment, that the first man, Adam, came 1,242 years before Noah (not reckoning by such Gentile calculations as Varro used, but simply by the Holy Scriptures, as best my weak understanding and astronomical calculations can interpret them.) About 1,080 years after Noah and the universal flood came Abraham, who, according to some, was a first-rate astrologer and invented the Chaldean alphabet. About 515 or 516 years later came Moses, and from his time to that of David about 570 years elapsed. From the time of David to that of out Savior and Redeemer, Jesus Christ, born of the unique Virgin, 1,350 year elapsed, according to some chronographs. Some may object that this calculation cannot be true, because it differs from that of Eusebius. From the time of the human redemption to the detestable heresy of the Saracens about 621 years elapsed. From this one can easily add up the amount of time gone by.

Although my calculations may not hold good for all nations, they have, however, been determined by the celestial movements, combined with the emotion, handed down to me by my forebears, which comes over me at certain hours. But the danger of the times, O Most Serene King, requires that such secrets should not be bared except in enigmatic sentences having, however, only one sense and meaning, and nothing ambiguous or amphibological inserted. Rather they are under a cloudy obscurity, with a natural infusion not unlike the creation of the world, according to the calculation and Punic Chronicle of Joel: I will pour out my spirit upon all flesh and your sons and daughters will prophesy. But such Prophecy proceeded from the mouth of the Holy Ghost who was the sovereign and eternal power, together with the heavens, and caused some of them to predict great and marvelous events.

As for myself, I would never claim such a title, never, please God. I readily admit that all proceeds from God and render to Him thanks, honor and immortal praise. I have mixed therewith no divination coming from fate. All from God and nature, and for the most part integrated with celestial movements. It is much like seeing in a burning mirror, with clouded vision, the great events, sad, prodigious and calamitous events that in due time will fall upon the principal worshippers. First, upon the temples of God; secondly, upon those who, sustained by the earth, approach such a decadence. Also a thousand other calamitous events which will be known to happen in due time.

For God will take notice of the long barrenness of the great dame, who thereupon will conceive two principal children. But she will be in danger, and the female to whom she will have given birth will also, because of the temerity of the age, be in danger of death in her eighteenth year, and will be unable to live beyond her thirty-sixth year. She will leave three males, and one female, and of these two will not have had the same father.

There will be great differences between the three brothers, and then there will be such great cooperation and agreement between them that the three and four parts of Europe will tremble. The youngest of them will sustain and augment the Christian monarchy, and under him sects will be elevated, and suddenly cast down, Arabs will be driven back, kingdoms united and new laws promulgated.

The oldest one will rule the land whose escutcheon is that of the furious crowned lions with their paws resting upon intrepid arms.

The one second in age, accompanied by the Latins, will penetrate far, until a second furious and trembling path has been beaten to the Great St. Bernard Pass. From there he will descend to mount the Pyrenees, which will not, however, be transferred to the French crown. And this third one will cause a great inundation of human blood, and for a long time Lent will not include March.

The daughter will be given for the preservation of the Christian Church. Her lord will fall into the pagan sect of the new infidels. Of her two children, one will be faithful to the Catholic Church, the other an infidel.

The unfaithful son, who, to his great confusion and later repentance, will want to ruin her, will have three widely scattered regions, namely, the Roman, Germany and Spain, which will set up diverse sects by armed force. The 50th to the 52th degree of latitude will be left behind.

And all will render the homage of ancient religions to the region of Europe north of the 48th parallel. The latter will have trembled first in vain timidity but afterwards the regions to its west, south and east will tremble. But the nature of their power will be such that what has been brought about by concord and union will prove insuperable by warlike conquests.

In nature they will be equal, but very different in faith.

After this the barren Dame, of greater power than the second, will be received by two of the nations. First, by them made obstinate by the onetime masters of the universe. Second, by the latter themselves.

The third people will extend their forces towards the circuit of the East of Europe where, in the Pannonias, they will be overwhelmed and slaughtered. By sea they will extend their Myrmidons and Germans to Adriatic Sicily. But they will succumb wholly and the Barbarian sect will be greatly afflicted and driven out by all the Latins.

Then the great Empire of the Antichrist will begin where once was Attila's empire and the new Xerxes will descend with great and countless numbers, so that the coming of the Holy Ghost, proceeding from the 48th degree, will make a transmigration, chasing out the abomination of the Christian Church, and whose reign will be for a time and to the end of time.

This will be preceded by a solar eclipse more dark and gloomy than any since the creation of the world, except that after the death and passion of Jesus Christ. And it will be in the month of October than the great translation will be made and it will be such that one will think the gravity of the earth has lost its natural movement and that it is to be plunged into the abyss of perpetual darkness.

In the spring there will be omens, and thereafter extreme changes, reversals of realms and mighty earthquakes. These will be accompanied by the procreation of the new Babylon, miserable daughter enlarged by the abomination of the first holocaust. It will last for only seventy-three years and seven months.

Then there will issue from the stock which had remained barren for so long, proceeding from the 50th degree, one who will renew the whole Christian Church. A great place will be established, with union and concord between some of the children of opposite ideas, who have been separated by diverse realms. And such will be the peace that the instigator and promoter of military factions, born of the diversity of religions, will remain chained to the deepest pit. And the kingdom of the Furious One, who counterfeits the sage, will be united.

The countries, towns, cities, realms and provinces which will have abandoned their old customs to gain liberty, but which will in fact have enthralled themselves even more, will secretly have wearied of their liberty. Faith lost in their perfect religion, they will begin to strike to the left, only to return to the right. Holiness, for a long time overcome, will be replaced in accordance with the earliest writings.

Thereafter the great dog, the biggest of curs, will go forth and destroy all, the same old crimes being perpetrated again. Temples will be set up again as in ancient times, and the priest will be restored to his original position and he will begin his whoring and luxury, and will commit a thousand crimes.

At the eve of another desolation, when she is atop her most high and sublime dignity, some potentates and warlords will confront her, and take away her two swords, and leave her only the insignia, whose curvature attracts them. The people will make him go to the right and will not wish to submit themselves to those of the opposite extreme with the hand in acute position, who touch the ground, and want to drive spurs into them.

The people of the world from this benevolent slavery to which they had voluntary submitted. He will put himself under the protection of Mars, stripping Jupiter of all his honors and dignities, and establish himself in the free city in another scant Mesopotamia. The chief and governor will be cast out from the middle and hung up, ignorant of the conspiracy of one of the conspirators with the second Thrasibulus, who for a long time will have directed all this.

Then the impurities and abominations, with a great shame, will be brought out and manifested in the shadows of the veiled light, and will cease towards the end of the change in reign. The chiefs of the Church will be backward in the love of God, and several of them will apostatize from the true faith. Of the three sects, that which is in the middle, because of its own partisans, will be thrown a bit into decadence. The first one will be exterminated throughout all Europe and most of Africa by the third one, making use of the poor in spirit who, led by madmen to libidinous luxury, will adulterate.

The supporting common people will rise up and chase out the adherents of the legislators. From the way realms will have been weakened by the Easterners, it will seem that God the Creator has loosed Satan from the prisons of hell to give birth to the great Dog and Dogam, who will make such an abominable breach in the Churches that neither the reds nor the whites without eyes or hands will know what to make of it, and their power will be taken from them.

Then will commence a persecution of the Churches the like of which was never seen. Meanwhile, such a plague will arise that more than two thirds of the world will be removed. One will be unable to ascertain the true owners of fields and houses, and weeds growing in the streets of cities will rise higher than the knees. For the clergy there will be but utter desolation. The warlords will usurp what is returned from the City of the Sun, from Malta and the Isles of Hyères. The great chain of the port which wakes its name from the marine ox will be opened.

And a new incursion will be made by the maritime shores, wishing to deliver the Sierra Morea from the first Mahometan recapture. Their assaults will not all be in vain, and the place which was once the abode of Abraham will be assaulted by persons who hold the Jovialists in veneration. And this city of "Achem" will be surrounded and assailed on all sides by a most powerful force of warriors. Their maritime forces will be weakened by the Westerners, and great desolation will fall upon this realm. Its greatest cities will be depopulated and those who enter will fall under the vengeance of the wrath of God.

The sepulcher, for long an object of such great veneration, will remain in the open, exposed to the sight of the heavens, the Sun and the Moon. The holy place will be converted into a stable for a herd large and small, and used for profane purposes. Oh, what a calamitous affliction will pregnant women bear at this time.

For hereupon the principal Eastern chief will be vanquished by the Northerners and Westerners, and most of his people, stirred up, will be put to death, overwhelmed or scattered. His children, offspring of many women, will be imprisoned. Then will be accomplished the prophecy of the Royal Prophet, Let him hear the groaning of the captives, that he might deliver the children of those doomed to die.

What great oppression will then fall upon the Princes and Governors of Kingdoms, especially those which will be maritime and Eastern, whose tongues will be intermingled with all others: the tongue of the Latins, and of the Arabs, via the Phoenicians. And all these Eastern Kings will be chased, overthrown and exterminated, but not altogether, by means of the forces of the Kings of the North, and because of the drawing near of our age through the three secretly united in the search for death, treacherously laying traps for one another. This renewed Triumvirate will last for seven years, and the renown of this sect will extend around the world. The sacrifice of the hole and immaculate Wafer will be sustained.

Then the Lords of "Aquilon" [the North], two in number, will be victorious over the Easterners, and so great a noise and bellicose tumult will they make amongst them that all the East will tremble in terror of these brothers, yet not brothers, of "Aquilon" [the North].

By this discourse, Sire, I present these predictions almost with confusion, especially as to when they will take place. Furthermore, the chronology of time which follows conforms very little, if at all, with that which has already been set forth. Yet it was determined by astronomy and other sources, including Holy Scriptures, and thus could not err. If I had wanted to date each quatrain, I could have done so. But this would not have been agreeable to all, least of all to those interpreting them, and was not to be done until Your Majesty granted me full power to do so, lest calumniators be furnished with an opportunity to injure me.

Anyhow, I count the years from the creation of the world to the birth of Noah as 1,506, and from the birth of Noah to the completion of the Ark, at the time of the universal deluge, as 600 (let the years be solar, or lunar, or a mixture of the ten) I hold that the Sacred Scriptures use solar years. And at the end of these 600 years, Noah entered the Ark to be saved from the deluge. This deluge was universal, and lasted one year and two months. And 295 years elapsed from the end of the flood to the birth of Abraham, and 100 from then till the birth of Isaac. And 60 years later Jacob was born. 130

years elapsed between the time he entered Egypt and the time he came out. Between the entry of Jacob into Egypt and the exodus, 430 years passed. From the exodus to the building of the Temple by Solomon in the fourth year of his reign, 480 years. According to the calculations of the Sacred Writings, it was 490 years from the building of the Temple to the time of Jesus Christ. Thus, this calculation of mine, collected from the holy writ, comes to about 4,173 years and 8 months, more or less. Because there is such a diversity of sects, I will not go beyond Jesus Christ.

I have calculated the present prophecies according to the order of the chain which contains its revolution, all by astronomical doctrine modified by my natural instinct. After a while, I found the time when Saturn turns to enter on April 7 till August 25, Jupiter on June 14 till October 7, Mars from April 17 to June 22, Venus from April 9 to May 22, Mercury from February 3 to February 24. After that, from June 1 to June 24, and from September 25 to October 16, Saturn in Capricorn, Jupiter in Aquarius, Mars in Scorpio, Venus in Pisces, Mercury for a month in Capricorn, Aquarius and Pisces, the Moon in Aquarius, the Dragon's head in Libra: its tail in opposition following a conjunction of Jupiter and Mercury with a quadrature of Mars and Mercury, and the Dragon's head coinciding with a conjunction of the Sun and Jupiter. And the year without an eclipse peaceful.

But not everywhere. It will mark the commencement of what will long endure. For beginning with this year the Christian Church will be persecuted more fiercely than it ever was in Africa, and this will last up to the year 1792, which they will believe to mark a renewal of time.

After this the Roman people will begin to re-establish themselves, chasing away some obscure shadows and recovering a bit of their ancient glory. But this will not be without great division and continual changes. Thereafter Venice will raise its wings very high in great force and power, not far short of the might of ancient Rome.

At that time the great sails of Byzantium, allied with the Ligurians and through the support and power of "Aquilon" [the Northern Realm], will impede them so greatly that the two Cretans will be unable to maintain their faith. The arks built by the Warriors of ancient times will accompany them to the waves of Neptune. In the Adriatic great discord will arise, and that which will have been united will be separated. To a house will be reduced that which was, and is, a great city, including "Pampotamia" and "Mesopotamia" of Europe at 45, and others of 41, 42 and 37 degrees.

It will be at this time and in these countries that the infernal power will set the power of its adversaries against the Church of Jesus Christ. This will constitute of the second Antichrist, who will persecute that Church and its true Vicar, by means of the power of three temporal kings who in their ignorance will be seduced by tongues which, in the hands of the madmen, will cut more than any sword.

The said reign of the Antichrist will last only to the death of him who was born at the beginning of the age and of the other one of Lyon, associated with the elected one of the House of Modena and of Ferrara, maintained by the Adriatic Ligurians and the proximity of great Sicily. Then the Great St. Bernard will be passed.

The Gallic Ogmios will be accompanied by so great a number that the Empire of his great law will extend very far. For some time thereafter the blood of the Innocent will be shed profusely by the recently elevated guilty ones. Then, because of great floods, the memory of things contained in these instruments will suffer incalculable loss, even letters. This will happen to the "Aquiloners" [the Northern People] by the will of God.

Once again Satan will be bound, universal peace will be established among men, and the Church of Jesus Christ will be delivered from all tribulations, although the Philistines would like to mix in the honey of malice and their pestilent seduction. This will be near the seventh millenary, when the sanctuary of Jesus Christ will no longer be trodden down by the infidels who come from "Aquilon" [the North]. The world will be approaching a great conflagration, although, according to my calculations in my prophecies, the course of time runs much further.

In the Epistle that some years ago I dedicated to my son, César Nostradamus, I declared some points openly enough, without presage. But here, Sire, are included several great and marvelous events which those to come after will see.

During this astrological supputation, harmonized with the Holy Scriptures, the persecution of the Ecclesiastical folk will have its origin in the power of the Kings of "Aquilon" [the North], united with the Easterners. This persecution will last for eleven years, or somewhat less, for then the chief King of "Aquilon" will fall.

Thereupon the same thing will occur in the South, where for the space of three years the Church people will be persecuted even more fiercely through the Apostatic seduction of one who will hold all the absolute power in the Church militant. The hole people of God, the observer of his law, will be persecuted fiercely and such will be their affliction that the blood of the true Ecclesiastics will flow everywhere.

One of the horrible temporal Kings will be told by his adherents, as the ultimate in praise, that he has shed more of human blood of Innocent Ecclesiastics than anyone else could have spilled of wine. This King will commit incredible crimes against the Church. Human blood will flow in the public streets and temples, like water after an impetuous rain, coloring the nearby rivers red with blood. The ocean itself will be reddened by another naval battle, such that one king will say to another, Naval battles have caused the sea to blush.

Then, in this same year, and in those following, there will ensue the most horrible pestilence, made more stupendous by the famine which will have preceded it. Such great tribulations will never have occurred since the first foundation of the Christian Church. It will cover all Latin regions, and will leave traces in some countries of the Spanish.

Thereupon the third King of "Aquilon" [the North], hearing the lament of the people of his principal title, will raise a very mighty army and, defying the tradition of his predecessors, will put almost everything back in its proper place, and the great Vicar of the hood will be put back in his former state. But desolated, and then abandoned by all, he will turn to find the Holy of Holies destroyed by paganism, and the old and new Testaments thrown out and burned.

After that Antichrist will be the infernal prince again, for the last time. All the Kingdoms of Christianity will tremble, even those of the infidels, for the space of twenty-five years. Wars and battles will be more grievous and towns, cities, castles and all other edifices will be burned, desolated and destroyed, with great effusion of vestal blood, violations of married woman and widows, and sucking children dashed and broken against the walls of towns. By means of Satan, Prince Infernal, so may evils will be committed that nearly all the world will find itself undone and desolated. Before these events, some rare birds will cry in the air: Hui, Hui [Today, today] and some time later will vanish.

After this has endured for a long time, there will be almost renewed another reign of Saturn, and golden age. Hearing the affliction of his people, God the Creator will command that Satan be cast into the depths of the bottomless pit, and bound there. Then a universal peace will commence between God and man, and Satan will remain bound for around a thousand years, and then all unbound.

All these figures represent the just integration of Holy Scriptures with visible celestial bodies, namely, Saturn, Jupiter, Mars and others conjoined, as can be seen at more length in some of the quatrains. I would have calculated more profoundly and integrated them even further, Most Serene King, but for the fact that some given to censure would raise difficulties. Therefore I withdraw my pen and seek nocturnal repose.

Many events, most powerful of all Kings, of the most astounding sort are to transpire soon, but I neither could nor would fit them all into this epistle; but in order to comprehend certain horrible facts, a few must be set forth. So great is your grandeur and humanity before men, and your piety before the gods, that you alone seem worthy of the great title of the Most Christian King, and to whom the highest authority in all religion should be deferred.

But I shall only beseech you, Most Clement King, by this singular and prudent humanity of yours, to understand rather the desire of my heart, and the sovereign wish I have to obey Your Most Serene Majesty, ever since my eyes approached your solar splendor, than the grandeur of my labor can attain to or acquire. From Salon, this 27th of June, 1558.

Done by Michel Nostradamus at Salon-de-Crau in Provence.

Century VIII

This is the eighth Century by Nostradamus. It was first published in 1568.

[Note: This is not my translation. It was found on the Internet a few years ago on a site that no longer exists. I have cleaned up gross misspellings and typos but have otherwise left the text intact. At some point, I hope to update with what I consider better translation. Due to lack of time on my part, this will have to suffice for now.]

For the French version, go to Nostradamiana at:
http://www.astrologer.ru:8003/Nostradamiana/centuries-eng.html

----

CENTURY VIII

1
Pau, Nay, Loron will be more of fire than blood,
to swim in praise, the great one to flee to the confluence (of rivers).
He will refuse entry to the magpies
Pampon and the Durance will keep them confined.

2
Condom and Auch and around Mirande,
I see fire from the sky which encompasses them.
Sun and Mars conjoined in Leo, then at Marmande,
lightning, great hail, a wall falls into the Garonne.

3
Within the strong castle of Vigilance and Resviers
the younger born of Nancy will be shut up.

In Turin the first ones will be burned,
when Lyons will be transported with grief.

4

The cock will be received into Monaco,
the Cardinal of France will appear;
He will be deceived by the Roman legation;
weakness to the eagle, strength will be born to the cock.

5

There will appear a shining ornate temple,
the lamp and the candle at Borne and Breteuil.
For the canton of Lucerne turned aside,
when one will see the great cock in his shroud.

6

Lighting and brightness are seen at Lyons shining,
Malta is taken, suddenly it will be extinguished.
Sardon, Maurice will act deceitfully,
Geneva to London, feigning treason towards the cock.

7

Vercelli, Milan will give the news,
the wound will be given at Pavia.
To run in the Seine, water, blood and fire through Florence,
the unique one falling from high to low calling for help.

8

Near Focia enclosed in some tuns
Chivasso will plot for the eagle.
The elected one driven out, he and his people shut up,
rape with Turin, the bride led away.

9

While the eagle is united with the cock at Savonna,
the Eastern Sea and Hungary.
The army at Naples, Palermo, the marches of Ancona,
Rome and Venice a great outcry by the Barbarian.

10

A great stench will come from Lausanne,
but they will not know its origin,
they will put out all people from distant places,
fire seen in the sky, a foreign nation defeated.

11

A multitude of people will appear at Vicenza
without force, fire to burn the Basilica.
Near Lunage the great one of Valenza defeated:
at a time when Venice takes up the quarrel through custom.

12
He will appear near to Buffalora
the highly born and tall one entered into Milan.
The Abbe of Foix with those of Saint-Meur
will cause damage dressed up as serfs.

13
The crusader brother through impassioned love
will cause Bellerophon to die through Proteus;
the fleet for a thousand years, the maddened woman,
the potion drunk, both of them then die.

14
The great credit of gold and abundance of silver
will cause honor to be blinded by lust;
the offense of the adulterer will become known,
which will occur to his great dishonor.

15
Great exertions towards the North by a man-woman
to vex Europe and almost all the Universe.
The two eclipses will be put into such a rout
that they will reinforce life or death for the Hungarians.

16
At the place where HIERON has his ship built,
there will be such a great sudden flood,
that one will not have a place nor land to fall upon,
the waters mount to the Olympic Fesulan.

17
Those at ease will suddenly be cast down,
the world put into trouble by three brothers;
their enemies will seize the marine city,
hunger, fire, blood, plague, all evils doubled.

18
The cause of her death will be issued from Florence,
one time before drunk by young and old;
by the three lilies they will give her a great pause.
Save through her offspring as raw meat is dampened.

19
To support the great troubled Cappe;
the reds will march in order to clarify it;
a family will be almost overcome by death,
the red, red ones will knock down the red one.

20
The false message about the rigged election
to run through the city stopping the broken pact;
voices bought, chapel stained with blood,
the empire contracted to another one.

21
Three foists will enter the port of Agde
carrying the infection and pestilence, not the faith.
Passing the bridge they will carry off a million,
the bridge is broken by the resistance of a third.

22
Coursan, Narbonne through the salt to warn
Tuchan, the grace of Perpignan betrayed;
the red town will not wish to consent to it,
in a high flight, a copy flag and a life ended.

23
Letters are found in the queen's chests,
no signature and no name of the author.
The ruse will conceal the offers;
so that they do not know who the lover is.

24
The lieutenant at the door of the house,
will knock down the great man of Perpignan.
Thinking to save himself at Montpertuis,
the bastard of Lusignan will be deceived.

25
The heart of the lover, awakened by furtive love
will ravish the lady in the stream.
She will pretend bashfully to be half injured,
the father of each will deprive the body of its soul.

26
The bones of Cato found in Barcelona,
placed, discovered, the site found again and ruined.

The great one who holds, but does not hold,
wants Pamplona, drizzle at the abbey of Montserrat.

27
The auxiliary way, one arch upon the other,
Le Muy deserted except for the brave one and his genet.
The writing of the Phoenix Emperor,
seen by him which is (shown) to no other.

28
The copies of gold and silver inflated,
which after the theft were thrown into the lake,
at the discovery that all is exhausted and dissipated by the debt.
All scrips and bonds will be wiped out.

29
At the fourth pillar which they dedicate to Saturn
split by earthquake and by flood;
under Saturn's building an urn is found
gold carried off by Caepio and then restored.

30
In Toulouse, not far from Beluzer
making a deep pit a palace of spectacle,
the treasure found will come to vex everyone
in two places and near the Basacle.

31
The first great fruit of the prince of Perchiera,
then will come a cruel and wicked man.
In Venice he will lose his proud glory,
and is led into evil by then younger Selin.

32
French king, beware of your nephew
who will do so much that your only son
will be murdered while making his vows to Venus;
accompanied at night by three and six.

33
The great one who will be born of Verona and Vincenza
who carries a very unworthy surname;
he who at Venice will wish to take vengeance,
himself taken by a man of the watch and sign.

34

After the victory of the Lion over the Lion,
there will be great slaughter on the mountain of Jura;
floods and dark-colored people of the seventh (of a million),
Lyons, Ulm at the mausoleum death and the tomb.

35
At the entrance to Garonne and Baise
and the forest not far from Damazan,
discoveries of the frozen sea, then hail and north winds.
Frost in the Dardonnais through the mistake of the month.

36
It will be committed against the anointed brought
from Lons le Saulnier, Saint Aubin and Bell'oeuvre.
To pave with marble taken from distant towers,
not to resist Bletteram and his masterpiece.

37
The fortress near the Thames
will fall when the king is locked up inside.
He will be seen in his shirt near the bridge,
one facing death then barred inside the fortress.

38
The King of Blois will reign in Avignon,
once again the people covered in blood.
In the Rhône he will make swim
near the walls up to five, the last one near Nolle.

39
He who will have been for the Byzantine prince
will be taken away by the prince of Toulouse.
The faith of Foix through the leader of Tolentino
will fail him, not refusing the bride.

40
The blood of the Just for Taur and La Duarade
in order to avenge itself against the Saturnines.
They will immerse the band in the new lake,
then they will march against Alba.

41
a fox will be elected without speaking one word,
appearing saintly in public living on barley bread,
afterwards he will suddenly become a tyrant
putting his foot on the throats of the greatest men.

42
Through avarice, through force and violence
the chief of Orléans will come to vex his supporters.
Near St. Memire, assault and resistance.
Dead in his tent they will say he is asleep inside.

43
Through the fall of two bastard creatures
the nephew of the blood will occupy the throne.
Within Lectoure there will be blows of lances,
the nephew through fear will fold up his standard.

44
The natural offspring off Ogmios
will turn off the road from seven to nine.
To the king long friend of the half man,
Navarre must destroy the fort at Pau.

45
With his hand in a sling and his leg bandaged,
the younger brother of Calais will reach far.
At the word of the watch, the death will be delayed,
then he will bleed at Easter in the Temple.

46
Paul the celibate will die three leagues from Rome,
the two nearest flee the oppressed monster.
When Mars will take up his horrible throne,
the Cock and the Eagle, France and the three brothers.

47
Lake Trasimene will bear witness
of the conspirators locked up inside Perugia.
A fool will imitate the wise one,
killing the Teutons, destroying and cutting to pieces.

48.
Saturn in Cancer, Jupiter with Mars
in February Chaldondon'salva tierra.
Sierra Morena besieged on three sides
near Verbiesque, war and mortal conflict.

49
Saturn in Taurus, Jupiter in Aquarius. Mars in Sagittarius,
the sixth of February brings death.

Those of Tardaigne so great a breach at Bruges,
that the barbarian chief will die at Ponteroso.

50
The plague around Capellades,
another famine is near to Sagunto;
the knightly bastard of the good old man
will cause the great one of Tunis to lose his head.

51
The Byzantine makes an oblation
after having taken back Cordoba.
A long rest on his road, the vines cut down,
at sea the passing prey captured by the Pillar.

52 ---- Unfinished/Censored ----
The king of Blois to reign in Avignon,
from Amboise and Seme the length of the Indre:
claws at Poitiers holy wings ruined
before Boni. . . .

53
Within Boulogne he will want to wash away his misdeeds,
he cannot at the temple of the Sun.
He will fly away, doing very great things:
In the hierarchy he had never an equal.

54
Under the color of the marriage treaty,
a magnanimous act by the Chyren Selin:
St. Quintin and Arras recovered on the journey;
By the Spanish a second butcher's bench is made.

55
He will find himself shut in between two rivers,
casks and barrels joined to cross beyond:
eight bridges broken, their chief run through so many times,
perfect children's throats slit by the knife.

56
The weak band will occupy the land,
those of high places will make dreadful cries.
The large herd of the outer corner troubled,
near Edinburgh it falls discovered by the writings.

57

From simple soldier he will attain to Empire,
from the short robe he will grow into the long.
Brave in arms, much worse towards the Church,
he vexes the priests as water fills a sponge.

58

A kingdom divided by two quarreling brothers
to take the arms and the name of Britain.
The Anglican title will be advised to watch out,
surprised by night (the other is), led to the French air.

59

Twice put up and twice cast down,
the East will also weaken the West.
Its adversary after several battles
chased by sea will fail at time of need.

60

First in Gaul, first in Romania,
over land and sea against the English and Paris.
Marvelous deeds by that great troop,
violent, the wild beast will lose Lorraine.

61

Never by the revelation of daylight
will he attain the mark of the scepter bearer.
Until all his sieges are at rest,
bringing to the Cock the gift of the armed legion.

62

When one sees the holy temple plundered,
the greatest of the Rhône profaning their sacred things;
because of them a very great pestilence will appear,
the king, unjust, will not condemn them.

63

When the adulterer wounded without a blow
will have murdered his wife and son out of spite;
his wife knocked down, he will strangle the child;
eight captives taken, choked beyond help.

64

The infants transported into the islands,
two out of seven will be in despair.
Those of the soil will be supported by it,
the name 'shovel' taken, the hope of the leagues fails.

65
The old man disappointed in his main hope,
will attain to the leadership of his Empire.
Twenty months he will hold rule with great force,
a tyrant, cruel, giving way to one worse.

66
When the inscription D.M. is found
in the ancient cave, revealed by a lamp.
Law, the King and Prince Ulpian tried,
the Queen and Duke in the pavilion under cover.

67
Paris, Carcassone, France to ruin in great disharmony,
neither one nor the other will be elected.
France will have the love and good will of the people,
Ferara, Colonna great protection.

68
The old Cardinal is deceived by the young one,
he will find himself disarmed, out of his position:
Do not show, Arles, that the double is perceived,
both Liqueduct and the Prince embalmed.

69
Beside the young one the old angel falls,
and will come to rise above him at the end;
ten years equal to most the old one falls again,
of three two and one, the eighth seraphim.

70
He will enter, wicked, unpleasant, infamous,
tyrannizing over Mesopotamia.
All friends made by the adulterous lady,
the land dreadful and black of aspect.

71
The number of astrologers will grow so great,
that they will be driven out, banned and their books censored.
In the year 1607 by sacred assemblies
so that none will be safe from the holy ones.

72
Oh what a huge defeat on the Perugian battlefield
and the conflict very close to Ravenna.

A holy passage when they will celebrate the feast,
the conqueror banished to eat horse meat.

73
The king is struck by a barbarian soldier,
unjustly, not far from death.
The greedy will be the cause of the deed,
conspirator and realm in great remorse.

74
A king entered very far into the new land
while the subjects will come to bid him welcome;
his treachery will have such a result
that to the citizens it is a reception instead of a festival.

75
The father and son will be murdered together,
the leader within his pavilion.
The mother at Tours will have her belly swollen with a son,
a verdure chest with little pieces of paper.

76
More of a butcher than a king in England,
born of obscure rank will gain empire through force.
Coward without faith, without law he will bleed the land;
His time approaches so close that I sigh.

77
The antichrist very soon annihilates the three,
twenty-seven years his war will last.
The unbelievers are dead, captive, exiled;
with blood, human bodies, water and red hail covering the earth.

78
A soldier of fortune with twisted tongue
will come to the sanctuary of the gods.
He will open the door to heretics
and raise up the Church militant.

79
He who loses his father by the sword, born in a Nunnery,
upon this Gorgon's blood will conceive anew;
in a strange land he will do everything to be silent,
he who will burn both himself and his child.

80

The blood of innocents, widow and virgin,
so many evils committed by means of the Great Red One,
holy images placed over burning candles,
terrified by fear, none will be seen to move.

81
The new empire in desolation
will be changed from the Northern Pole.
From Sicily will come such trouble that
it will bother the enterprise tributary to Philip.

82
Thin tall and dry, playing the good valet
in the end will have nothing but his dismissal;
sharp poison and letters in his collar,
he will be seized escaping into danger.

83
The largest sail set out of the port of Zara,
near Byzantium will carry out its enterprise.
Loss of enemy and friend will not be,
a third will turn on both with great pillage and capture.

84
Paterno will hear the cry from Sicily,
all the preparations in the Gulf of Trieste;
it will be heard as far as Sicily
flee oh, flee, so may sails, the dreaded pestilence !

85
Between Bayonne and St. Jean de Luz
will be placed the promontory of Mars.
To the Hanix of the North, Nanar will remove the light,
then suffocate in bed without assistance.

86
Through Emani, Tolosa and Villefranche,
an infinite band through the mountains of Adrian.
Passes the river, Cambat over the plank for a bridge,
Bayonne will be entered all crying Bigoree.

87
A death conspired will come to its full effect,
the charge given and the voyage of death.
Elected, created, received (then) defeated by its followers,
in remorse the blood of innocence in front of him.

88

A noble king will come to Sardinia,
who will only rule for three years in the kingdom.
He will join with himself several colors;
he himself, after taunts, care spoils slumber.

89

In order not to fall into the hands of his uncle
who slaughtered his children in order to reign.
Pleasing with the people, putting his foot on Peloncle,
dead and dragged between armored horses.

90

When those of the cross are found their senses troubled,
in place of sacred things he will see a horned bull,
through the virgin the pig's place will then be filled,
order will no longer be maintained by the king.

91

Entered among the field of the Rhône
where those of the cross are almost united,
the two lands meeting in Pisces
and a great number punished by the flood.

92

Far distant from his kingdom, sent on a dangerous journey,
he will lead a great army and keep it for himself.
The king will hold his people captive and hostage,
he will plunder the whole country on his return.

93

For seven months, no longer, will he hold the office of prelate,
through his death a great schism will arise;
for seven months another acts as prelate near Venice,
peace and union are reborn.

94

In front of the lake where the dearest one was destroyed
for seven months and his army routed;
Spaniards will be devastating by means of Alba,
through delay in giving battle, loss.

95

The seducer will be placed in a ditch
and will be tied up for some time.

The scholar joins the chief with his cross.
The sharp right will draw the contented ones.

96
The sterile synagogue without any fruit,
will be received by the infidels,
the daughter of the persecuted (man) of Babylon,
miserable and sad, they will clip her wings.

97
At the end of the Var the great powers change;
near the bank three beautiful children are born.
Ruin to the people when they are of age;
in the country the kingdom is seen to grow and change more.

98
Of the church men the blood will be poured forth
as abundant as water in (amount);
for a long time it will not be restrained,
woe, woe, for the clergy ruin and grief.

99
Through the power of three temporal kings,
the sacred seat will be put in another place,
where the substance of the body and the spirit
will be restored and received as the true seat.

100
By the great number of tears shed,
from top to bottom and from the bottom to the very top,
a life is lost through a game with too much faith,
to die of thirst through a great deficiency.

Century IX

This is the ninth Century by Nostradamus. It was first published in 1568.

[Note: This is not my translation. It was found on the Internet a few years ago on a site that no longer exists. I have cleaned up gross misspellings and typos but have otherwise left the text intact. At some point, I hope to update with what I consider better translation. Due to lack of time on my part, this will have to suffice for now.]

For the French version, go to Nostradamiana at:
http://www.astrologer.ru:8003/Nostradamiana/centuries-eng.html

CENTURY IX

1
In the house of the translator of Bourg,
The letters will be found on the table,
One-eyed, red-haired, white, hoary-headed will hold the course,
Which will change for the new Constable.

2
From the top of the Aventine hill a voice heard,
Be gone, be gone all of you on both sides:
The anger will be appeased by the blood of the red ones,
From Rimini and Prato, the Colonna expelled.

3
The "great cow" at Racenna in great trouble,
Led by fifteen shut up at Fornase:
At Rome there will be born two double-headed monsters,
Blood, fire, flood, the greatest ones in space.

4
The following year discoveries through flood,
Two chiefs elected, the first one will not hold:
The refuge for the one of them fleeing a shadow,
The house of which will maintain the first one plundered.

5
The third toe will seem first
To a new monarch from low high,
He who will possess himself as a Tyrant of Pisa and Lucca,
To correct the fault of his predecessor.

6
An infinity of Englishmen in Guienne
Will settle under the name of Anglaquitaine:
In Languedoc, Ispalme, Bordelais,
Which they will name after Barboxitaine.

7
He who will open the tomb found,
And will come to close it promptly,
Evil will come to him, and one will be unable to prove,
If it would be better to be a Breton or Norman King.

8
The younger son made King will put his father to death,
After the conflict very dishonest death:
Inscription found, suspicion will bring remorse,
When the wolf driven out lies down ion the bedstead.

9
When the lamp burning with inextinguishable fire
Will be found in the temple of the Vestals:
Child found in fire, water passing through the sieve:
To perish in water Nîmes, Toulouse the markets to fall.

10
The child of a monk and nun exposed to death,
To die through a she-bear, and carried off by a boar,
The army will be camped by Foix and Pamiers,
Against Toulouse Carcassonne the harbinger to form.

11
Wrongly will they come to put the just one to death,
In public and in the middle extinguished:
So great a pestilence will come to arise in this place,
That the judges will be forced to flee.

12
So much silver of Diana and Mercury,
The images will be found in the lake:
The sculptor looking for new clay,
He and his followers will be steeped in gold.

13
The exiles around Sologne,
Led by night to march into Auxois,
Two of Modena for Bologna cruel,
Placed discovered by the fire of Buzanais.

14
Dyers' caldrons put on the flat surface,
Wine, honey and oil, and built over furnaces:
They will be immersed, innocent, pronounced malefactors,
Seven of Bordeaux smoke still in the cannon.

15
Near Perpignan the red ones detained,
Those of the middle completely ruined led far off:
Three cut in pieces, and five badly supported,

For the Lord and Prelate of Burgundy.

16
Out of Castelfranco will come the assembly,
The ambassador not agreeable will cause a schism:
Those of Riviera will be in the squabble,
And they will refuse entry to the great gulf.

17
The third one first does worse than Nero,
How much human blood to flow, valiant, be gone:
He will cause the furnace to be rebuilt,
Golden Age dead, new King great scandal.

18
The lily of the Dauphin will reach into Nancy,
As far as Flanders the Elector of the Empire:
New confinement for the great Montmorency,
Outside proven places delivered to celebrated punishment.

19
In the middle of the forest of Mayenne,
Lightning will fall, the Sun in Leo:
The great bastard issued from the great one Maine,
On this day a point will enter the blood of Fougères.

20
By night will come through the forest of Reines,
Two couples roundabout route Queen the white stone,
The monk king in gray in Varennes:
Elected Capet causes tempest, fire, blood, slice.

21
At the tall temple of Saint-Solenne at Blois,
Night Loire bridge, Prelate, King killing outright:
Crushing victory in the marshes of the pond,
Whence prelacy of whites miscarrying.

22
The King and his court in the place of cunning tongue,
Within the temple facing the palace:
In the garden the Duke of Mantua and Alba,
Alba and Mantua dagger tongue and palace.

23
The younger son playing outdoors under the arbor,

The top of the roof in the middle on his head,
The father King in the temple of Saint-Solonne,
Sacrificing he will consecrate festival smoke.

24
Upon the palace at the balcony of the windows,
The two little royal ones will be carried off:
To pass Orléans, Paris, abbey of Saint-Denis,
Nun, wicked ones to swallow green pits.

25
Crossing the bridges to come near the Roisiers,
Sooner than he thought, he arrived late.
The new Spaniards will come to Béziers,
So that this chase will break the enterprise.

26
Departed by the bitter letters the surname of Nice,
The great Cappe will present something, not his own;
Near Voltai at the wall of the green columns,
After Piombino the wind in good earnest.

27
The forester, the wind will be close around the bridge,
Received highly, he will strike the Dauphin.
The old craftsman will pass through the woods in a company,
Going far beyond the right borders of the Duke.

28
The Allied fleet from the port of Marseilles,
In Venice harbor to march against Hungary.
To leave from the gulf and the bay of Illyria,
Devastation in Sicily, for the Ligurians, cannon shot.

29
When the man will give way to none,
Will wish to abandon a place taken, yet not taken;
Ship afire through the swamps, bitumen at Charlieu,
St. Quintin and Calais will be recaptured.

30
At the port of Pola and of San Nicolo,
A Normand will punish in the Gulf of Quarnero:
Capet to cry alas in the streets of Byzantium,
Help from Cadiz and the great Philip.

31
The tin island of St. George half sunk;
Drowsy with peace, war will arise,
At Easter in the temple abysses opened.

32
A deep column of fine porphyry is found,
Inscriptions of the Capitol under the base;
Bones, twisted hair, the Roman strength tried,
The fleet is stirred at the harbor of Mitylene.

33
Hercules King of Rome and of "Annemark,"
With the surname of the chief of triple Gaul,
Italy and the one of St. Mark to tremble,
First monarch renowned above all.

34
The single part afflicted will be mitered,
Return conflict to pass over the tile:
For five hundred one to betray will be titled
Narbonne and Salces we have oil for knives.

35
And fair Ferdinand will be detached,
To abandon the flower, to follow the Macedonian:
In the great pinch his course will fail,
And he will march against the Myrmidons.

36
A great King taken by the hands of a young man,
Not far from Easter confusion knife thrust:
Everlasting captive times what lightning on the top,
When three brothers will wound each other and murder.

37
Bridge and mills overturned in December,
The Garonne will rise to a very high place:
Walls, edifices, Toulouse overturned,
So that none will know his place like a matron.

38
The entry at Blaye for La Rochelle and the English,
The great Macedonian will pass beyond:
Not far from Agen will wait the Gaul,
Narbonne help beguiled through conversation.

39

In Albisola to Veront and Carcara,
Led by night to seize Savona:
The quick Gascon La Turbie and L'Escarène:
Behind the wall old and new palace to seize.

40

Near Saint-Quintin in the forest deceived,
In the Abbey the Flemish will be cut up:
The two younger sons half-stunned by blows,
The rest crushed and the guard all cut to pieces.

41

The great "Chyren" will seize Avignon,
From Rome letters in honey full of bitterness:
Letter and embassy to leave from Chanignon,
Carpentras taken by a black duke with a red feather.

42

From Barcelona, from Genoa and Venice,
From Sicily pestilence Monaco joined:
They will take their aim against the Barbarian fleet,
Barbarian driven 'way back as far as Tunis.

43

On the point of landing the Crusader army
Will be ambushed by the Ishmaelites,
Struck from all sides by the ship Impetuosity,
Rapidly attacked by ten elite galleys.

44

Leave, leave Geneva every last one of you,
Saturn will be converted from gold to iron,
Raypoz will exterminate all who oppose him,
Before the coming the sky will show signs.

45

None will remain to ask,
Great Mendosus will obtain his dominion:
Far from the court he will cause to be countermanded
Piedmont, Picardy, Paris, Tuscany the worst.

46

Be gone, flee from Toulouse ye red ones,
For the sacrifice to make expiation:

The chief cause of the evil under the shade of pumpkins:
Dead to strangle carnal prognostication.

47
The undersigned to an infamous deliverance,
And having contrary advice from the multitude:
Monarch changes put in danger over thought,
Shut up in a cage they will see each other face to face.

48
The great city of the maritime Ocean,
Surrounded by a crystalline swamp:
In the winter solstice and the spring,
It will be tried by frightful wind.

49
Ghent and Brussels will march against Antwerp,
The Senate of London will put to death their King:
Salt and wine will overthrow him,
To have them the realm turned upside down.

50
Mendosus will soon come to his high realm,
Putting behind a little the Lorrainers:
The pale red one, the male in the interregnum,
The fearful youth and Barbaric terror.

51
Against the red ones sects will conspire,
Fire, water, steel, rope through peace will weaken:
On the point of dying those who will plot,
Except one who above all the world will ruin.

52
Peace is nigh on one side, and war,
Never was the pursuit of it so great:
To bemoan men, women innocent blood on the land,
And this will be throughout all France.

53
The young Nero in the three chimneys
Will cause live pages to be thrown to burn:
Happy those who will be far away from such practices,
Three of his blood will have him ambushed to death.

54

There will arrive at Porto Corsini,
Near Ravenna, he who will plunder the lady:
In the deep sea legate from Lisbon,
Hidden under a rock they will carry off seventy souls.

55
The horrible war which is being prepared in the West,
The following year will come the pestilence
So very horrible that young, old, nor beast,
Blood, fire Mercury, Mars, Jupiter in France.

56
The army near Houdan will pass Goussainville,
And at Maiotes it will leave its mark:
In an instant more than a thousand will be converted,
Looking for the two to put them back in chain and firewood.

57
In the place of Drux a King will rest,
And will look for a law changing Anathema:
While the sky will thunder so very loudly,
New entry the King will kill himself.

58
On the left side at the spot of Vitry,
The three red ones of France will be awaited:
All felled red, black one not murdered,
By the Bretons restored to safety.

59
At La Ferté-Vidame he will seize,
Nicholas held red who had produced his life:
The great Louise who will act secretly one will be born,
Giving Burgundy to the Bretons through envy.

60
Conflict Barbarian in the black Headdress,
Blood shed, Dalmatia to tremble:
Great Ishmael will set up his promontory,
Frogs to tremble Lusitania aid.

61
The plunder made upon the marine coast,
In Cittanova and relatives brought forward:
Several of Malta through the deed of Messina
Will be closely confined poorly rewarded.

62
To the great one of Ceramon-agora,
The crusaders will all be attached by rank,
The long-lasting Opium and Mandrake,
The Raugon will be released on the third of October.

63
Complaints and tears, cries and great howls,
Near Narbonne at Bayonne and in Foix:
Oh, what horrible calamities and changes,
Before Mars has made several revolutions.

64
The Macedonian to pass the Pyrenees mountains,
In March Narbonne will not offer resistance:
By land and sea he will carry on very great intrigue,
Capetian having no land safe for residence.

65
He will come to go into the corner of Luna,
Where he will be captured and put in a strange land:
The unripe fruits will be the subject of great scandal,
Great blame, to one great praise.

66
There will be peace, union and change,
Estates, offices, low high and high very low:
To prepare a trip, the first offspring torment,
War to cease, civil process, debates.

67
From the height of the mountains around the Isère,
One hundred assembled at the haven in the rock Valence:
From Châteauneuf, Pierrelatte, in Donzère,
Against Crest, Romans, faith assembled.

68
The noble of Mount Aymar will be made obscure,
The evil will come at the junction of the Saône and Rhône:
Soldiers hidden in the woods on Lucy's day,
Never was there so horrible a throne.

69
One the mountain of Saint-Bel and L'Arbresle
The proud one of Grenoble will be hidden:

Beyond Lyons and Vienne on them a very great hail,
Lobster on the land not a third thereof will remain.

70

Sharp weapons hidden in the torches.
In Lyons, the day of the Sacrament,
Those of Vienne will all be cut to pieces,
By the Latin Cantons Mâcon does not lie.

71

At the holy places animals seen with hair,
With him who will not dare the day:
At Carcassonne propitious for disgrace,
He will be set for a more ample stay.

72

Again will the holy temples be polluted,
And plundered by the Senate of Toulouse:
Saturn two three cycles completed,
In April, May, people of new leaven.

73

The Blue Turban King entered into Foix,
And he will reign less than an evolution of Saturn:
The White Turban King Byzantium heart banished,
Sun, Mars and Mercury near Aquarius.

74

In the city of Fertsod homicide,
Deed, and deed many oxen plowing no sacrifice:
Return again to the honors of Artemis,
And to Vulcan bodies dead ones to bury.

75

From Ambracia and the country of Thrace
People by sea, evil and help from the Gauls:
In Provence the perpetual trace,
With vestiges of their custom and laws.

76

With the rapacious and blood-thirsty king,
Issued from the pallet of the inhuman Nero:
Between two rivers military hand left,
He will be murdered by Young Baldy.

77

The realm taken the King will conspire,
The lady taken to death ones sworn by lot:
They will refuse life to the Queen and son,
And the mistress at the fort of the wife.

78
The Greek lady of ugly beauty,
Made happy by countless suitors:
Transferred out to the Spanish realm,
Taken captive to die a miserable death.

79
The chief of the fleet through deceit and trickery
Will make the timid ones come out of their galleys:
Come out, murdered, the chief renouncer of chrism,
Then through ambush they will pay him his wages.

80
The Duke will want to exterminate his followers,
He will send the strongest ones to strange places:
Through tyranny to ruin Pisa and Lucca,
Then the Barbarians will gather the grapes without vine.

81
The crafty King will understand his snares,
Enemies to assail from three sides:
A strange number tears from hoods,
The grandeur of the translator will come to fail.

82
By the flood and fierce pestilence,
The great city for long besieged:
The sentry and guard dead by hand,
Sudden capture but none wronged.

83
Sun twentieth of Taurus the earth will tremble very mightily,
It will ruin the great theater filled:
To darken and trouble air, sky and land,
Then the infidel will call upon God and saints.

84
The King exposed will complete the slaughter,
After having discovered his origin:
Torrent to open the tomb of marble and lead,
Of a great Roman with Medusine device.

85

To pass Guienne, Languedoc and the Rhône,
From Agen holding Marmande and La Réole:
To open through faith the wall, Marseilles will hold its throne,
Conflict near Saint-Paul-de-Mausole.

86

From Bourg-la-Reine they will come straight to Chartres,
And near Pont d'Antony they will pause:
Seven crafty as Martens for peace,
Paris closed by an army they will enter.

87

In the forest cleared of the Tuft,
By the hermitage will be placed the temple:
The Duke of Étampes through the ruse he invented
Will teach a lesson to the prelate of Montlhéry.

88

Calais, Arras, help to Thérouanne,
Peace and semblance the spy will simulate:
The soldiery of Savoy to descend by Roanne,
People who would end the rout deterred.

89

For seven years fortune will favor Philip,
He will beat down again the exertions of the Arabs:
Then at his noon perplexing contrary affair,
Young Ogmios will destroy his stronghold.

90

A captain of Great Germany
Will come to deliver through false help
To the King of Kings the support of Pannonia,
So that his revolt will cause a great flow of blood.

91

The horrible plague Perinthus and Nicopolis,
The Peninsula and Macedonia will it fall upon:
It will devastate Thessaly and Amphipolis,
An unknown evil, and from Anthony refusal.

92

The King will want to enter the new city,
Through its enemies they will come to subdue it:

Captive free falsely to speak and act,
King to be outside, he will keep far from the enemy.

93

The enemies very far from the fort,
The bastion brought by wagons:
Above the walls of Bourges crumbled,
When Hercules the Macedonian will strike.

94

Weak galleys will be joined together,
False enemies the strongest on the rampart:
Weak ones assailed Bratislava trembles,
Lübeck and Meissen will take the barbarian side.

95

The newly made one will lead the army,
Almost cut off up to near the bank:
Help from the Milanais elite straining,
The Duke deprived of his eyes in Milan in an iron cage.

96

The army denied entry to the city,
The Duke will enter through persuasion:
The army led secretly to the weak gates,
They will put it to fire and sword, effusion of blood.

97

The forces of the sea divided into three parts,
The second one will run out of supplies,
In despair looking for the Elysian Fields,
The first ones to enter the breach will obtain the victory.

98

Those afflicted through the fault of a single one stained,
The transgressor in the opposite party:
He will send word to those of Lyons that compelled
They be to deliver the great chief of Molite.

99

The "Aquilon" Wind will cause the siege to be raised,
Over the walls to throw ashes, lime and dust:
Through rain afterwards, which will do them much worse,
Last help against their frontier.

100

Naval battle night will be overcome,
Fire in the ships to the West ruin:
New trick, the great ship colored,
Anger to the vanquished, and victory in a drizzle.

Century X

This is the tenth and final Century of Nostradamus.

[Note: This is not my translation. It was found on the Internet a few years ago on a site that no longer exists. I have cleaned up gross misspellings and typos but have otherwise left the text intact. At some point, I hope to update with what I consider better translation. Due to lack of time on my part, this will have to suffice for now.]

For the French version, go to Nostradamiana at:
http://www.astrologer.ru:8003/Nostradamiana/centuries-eng.html

----
CENTURY X

1
To the enemy, the enemy faith promised
Will not be kept, the captives retained:
One near death captured, and the remainder in their shirts,
The remainder damned for being supported.

2
The ship's veil will hide the sail galley,
The great fleet will come the lesser one to go out:
Ten ships near will turn to drive it back,
The great one conquered the united ones to join to itself.

3
After that five will not put out the flock,
A fugitive for Penelon he will turn loose:
To murmur falsely then help to come,
The chief will then abandon the siege.

4
At midnight the leader of the army
Will save himself, suddenly vanished:
Seven years later his reputation unblemished,
To his return they will never say yes.

5
Albi and Castres will form a new league,
Nine Arians Lisbon and the Portuguese:
Carcassonne and Toulouse will end their intrigue,
When the chief new monster from the Lauraguais.

6
The Gardon will flood Nîmes so high
That they will believe Deucalion reborn:
Into the colossus the greater part will flee,
Vesta tomb fire to appear extinguished.

7
The great conflict that they are preparing for Nancy,
The Macedonian will say I subjugate all:
The British Isle in anxiety over wine and salt,
"Hem. mi." Philip two Metz will not hold for long.

8
With forefinger and thumb he will moisten the forehead,
The Count of Senigallia to his own son:
The Venus through several of thin forehead,
Three in seven days wounded dead.

9
In the Castle of Figueras on a misty day
A sovereign prince will be born of an infamous woman:
Surname of breeches on the ground will make him posthumous,
Never was there a King so very bad in his province.

10
Stained with murder and enormous adulteries,
Great enemy of the entire human race:
One who will be worse than his grandfathers, uncles or fathers,
In steel, fire, waters, bloody and inhuman.

11
At the dangerous passage below Junquera,
The posthumous one will have his band cross:
To pass the Pyrenees mountains without his baggage,
From Perpignan the duke will hasten to Tende.

12
Elected Pope, as elected he will be mocked,
Suddenly unexpectedly moved prompt and timid:
Through too much goodness and kindness provoked to die,

Fear extinguished guides the night of his death.

13
Beneath the food of ruminating animals,
led by them to the belly of the fodder city:
Soldiers hidden, their arms making a noise,
Tried not far from the city of Antibes.

14
Urnel Vaucile without a purpose on his own,
Bold, timid, through fear overcome and captured:
Accompanied by several pale whores,
Convinced in the Carthusian convent at Barcelona.

15
Father duke old in years and choked by thirst,
On his last day his don denying him the jug:
Into the well plunged alive he will come up dead,
Senate to the thread death long and light.

16
Happy in the realm of France, happy in life,
Ignorant of blood, death, fury and plunder:
For a flattering name he will be envied,
A concealed King, too much faith in the kitchen.

17
The convict Queen seeing her daughter pale,
Because of a sorrow locked up in her breast:
Lamentable cries will come then from Angoulême,
And the marriage of the first cousin impeded.

18
The house of Lorraine will make way for Vendôme,
The high put low, and the low put high:
The son of Mammon will be elected in Rome,
And the two great ones will be put at a loss.

29
The day that she will be hailed as Queen,
The day after the benediction the prayer:
The reckoning is right and valid,
Once humble never was one so proud.

20
All the friend who will have belonged to the party,

For the rude in letters put to death and plundered:
Property up for sale at fixed price the great one annihilated.
Never were the Roman people so wronged.

21
Through the spite of the King supporting the lesser one,
He will be murdered presenting the jewels to him:
The father wishing to impress nobility on the son
Does as the Magi did of yore in Persia.

22
For not wishing to consent to the divorce,
Which then afterwards will be recognized as unworthy:
The King of the Isles will be driven out by force,
In his place put one who will have no mark of a king.

23
The remonstrances made to the ungrateful people,
Thereupon the army will seize Antibes:
The complaints will place Monace in the arch,
And at Fréjus the one will take the shore from the other

24
The captive prince conquered in Italy
Will pass Genoa by sea as far as Marseilles:
Through great exertion by the foreigners overcome,
Safe from gunshot, barrel of bee's liquor.

25
Through the Ebro to open the passage of Bisanne,
Very far away will the Tagus make a demonstration:
In Pelligouxe will the outrage be committed,
By the great lady seated in the orchestra.

26
The successor will avenge his brother-in-law,
To occupy the realm under the shadow of vengeance:
Obstacle slain his blood for the death blame,
For a long time will Brittany hold with France.

27
Through the fifth one and a great Hercules
They will come to open the temple by hand of war:
One Clement, Julius and Ascanius set back,
The sword, key, eagle, never was there such a great animosity.

28
Second and third which make prime music
By the King to be sublimated in honor:
Through the fat and the thin almost emaciated,
By the false report of Venus to be debased.

29
In a cave of Saint-Paul-de-Mausole a goat
Hidden and seized pulled out by the beard:
Led captive like a mastiff beast
By the Bigorre people brought to near Tarbes.

30
Nephew and blood of the new saint come,
Through the surname he will sustain arches and roof:
They will be driven out put to death chased nude,
Into red and black will they convert their green.

31
The Holy Empire will come into Germany,
The Ishmaelites will find open places:
The asses will want also Carmania,
The supporters all covered by earth.

32
The great empire, everyone would be of it,
One will come to obtain it over the others:
But his realm and state will be of short duration,
Two years will he be able to maintain himself on the sea.

33
The cruel faction in the long robe
Will come to hide under the sharp daggers:
The Duke to seize Florence and the diphthong place,
Its discovery by immature ones and sycophants.

34
The Gaul who will hold the empire through war,
He will be betrayed by his minor brother-in-law:
He will be drawn by a fierce, prancing horse,
The brother will be hated for the deed for a long time

35
The younger son of the king flagrant in burning lust
To enjoy his first cousin:
Female attire in the Temple of Artemis,

Going to be murdered by the unknown one of Maine.

36
Upon the King of the stump speaking of wars,
The United Isle will hold him in contempt:
For several good years one gnawing and pillaging,
Through tyranny in the isle esteem changing.

37
The great assembly near the Lake of Bourget,
They will meet near Montmélian:
Going beyond the thoughtful ones will draw up a plan,
Chambéry, Saint-Jean-de-Maurienne, Saint-Julien combat.

38
Sprightly love lays the siege not far,
The garrisons will be at the barbarian saint:
The Orsini and Adria will provide a guarantee for the Gauls,
For fear delivered by the army to the Grisons.

39
First son, widow, unfortunate marriage,
Without any children two Isles in discord:
Before eighteen, incompetent age,
For the other one the betrothal will take place while younger.

40
The young heir to the British realm,
Whom his dying father will have recommended:
The latter dead Lonole will dispute with him,
And from the son the realm demanded.

41
On the boundary of Caussade and Caylus,
Not at all far from the bottom of the valley:
Music from Villefranche to the sound of lutes,
Encompassed by cymbals and great stringing.

42
The humane realm of Anglican offspring,
It will cause its realm to hold to peace and union:
War half-captive in its enclosure,
For long will it cause them to maintain peace.

43
Too much good times, too much of royal goodness,

Ones made and unmade, quick, sudden, neglectful:
Lightly will he believe falsely of his loyal wife,
He put to death through his benevolence.

44
When a King will be against his people,
A native of Blois will subjugate the Ligurians,
Memel, Cordoba and the Dalmatians,
Of the seven then the shadow to the King, New Year's money and ghosts.

45
The shadow of the realm of Navarre untrue,
It will make his life one of fate unlawful:
The vow made in Cambrai wavering,
King Orléans will give a lawful wall.

46
In life, fate and death a sordid, unworthy man of gold,
He will not be a new Elector of Saxony:
From Brunswick he will send for a sign of love,
The false seducer delivering it to the people.

47
At the Garland lady of the town of Burgos,
They will impose for the treason committed:
The great prelate of Leon through Formande,
Undone by false pilgrims and ravishers.

48
Banners of the deepest part of Spain,
Coming out from the tip and ends of Europe:
Troubles passing near the bridge of Laigne,
Its great army will be routed by a band.

49
Garden of the world near the new city,
In the path of the hollow mountains:
It will be seized and plunged into the Tub,
Forced to drink waters poisoned by sulfur.

50
The Meuse by day in the land of Luxembourg,
It will find Saturn and three in the urn:
Mountain and plain, town, city and borough,
Flood in Lorraine, betrayed by the great urn.

51
Some of the lowest places of the land of Lorraine
Will be united with the Low Germans:
Through those of the see Picards, Normans, those of Main,
And they will be joined to the cantons.

52
At the place where the Lys and the Scheldt unite,
The nuptials will be arranged for a long time:
At the place in Antwerp where they carry the chaff,
Young old age wife undefiled.

53
The three concubines will fight each other for a long time,
The greatest one the least will remain to watch:
The great Selin will no longer be her patron,
She will call him fire shield white route.

54
She born in this world of a furtive concubine,
At two raised high by the sad news:
She will be taken captive by her enemies,
And brought to Malines and Brussels.

55
The unfortunate nuptials will be celebrated
In great joy but the end unhappy:
Husband and mother will slight the daughter-in-law,
The Apollo dead and the daughter-in-law more pitiful.

56
The royal prelate his bowing too low,
A great flow of blood will come out of his mouth:
The Anglican realm a realm pulled out of danger,
For long dead as a stump alive in Tunis.

57
The uplifted one will not know his scepter,
He will disgrace the young children of the greatest ones:
Never was there a more filthy and cruel being,
For their wives the king will banish them to death.

58
In the time of mourning the feline monarch
Will make war upon the young Macedonian:
Gaul to shake, the bark to be in jeopardy,

Marseilles to be tried in the West a talk.

59
Within Lyons twenty-five of one mind,
Five citizens, Germans, Bressans, Latins:
Under a noble one they will lead a long train,
And discovered by barks of mastiffs.

60
I weep for Nice, Monaco, Pisa, Genoa,
Savona, Siena, Capua, Modena, Malta:
For the above blood and sword for a New Year's gift,
Fire, the earth will tremble, water an unhappy reluctance.

61
Betta, Vienna, Emorte, Sopron,
They will want to deliver Pannonia to the Barbarians:
Enormous violence through pike and fire,
The conspirators discovered by a matron.

62
Near "Sorbia" to assail Hungary,
The herald of "Brudes" (dark ones?) will come to warn them:
Byzantine chief, Salona of Slavonia,
He will come to convert them to the law of the Arabs.

63
Cydonia, Ragusa, the city of St. Jerome,
With healing help to grow green again:
The King's son dead because of the death of two heroes,
Araby and Hungary will take the same course.

64
Weep Milan, weep Lucca and Florence,
As your great Duke climbs into the chariot:
The see to change it advances to near Venice,
When at Rome the Colonna will change.

65
O vast Rome, thy ruin approaches,
Not of thy walls, of thy blood and substance:
The one harsh in letters will make a very horrible notch,
Pointed steel driven into all up to the hilt.

66
The chief of London through the realm of America,

The Isle of Scotland will be tried by frost:
King and Reb will face an Antichrist so false,
That he will place them in the conflict all together.

67
A very mighty trembling in the month of May,
Saturn in Capricorn, Jupiter and Mercury in Taurus:
Venus also, Cancer, Mars in Virgo,
Hail will fall larger than an egg.

68
The army of the sea will stand before the city,
Then it will leave without making a long passage:
A great flock of citizens will be seized on land,
Fleet to return to seize it great robbery.

69
The shining deed of the old one exalted anew,
Through the South and Aquilon they will be very great:
Raised by his own sister great crowds,
Fleeing, murdered in the thicket of Ambellon.

70
Through an object the eye will swell very much,
Burning so much that the snow will fall:
The fields watered will come to shrink,
As the primate succumbs at Reggio.

71
The earth and air will freeze a very great sea,
When they will come to venerate Thursday:
That which will be never was it so fair,
From the four parts they will come to honor it.

72
The year 1999, seventh month,
From the sky will come a great King of Terror:
To bring back to life the great King of the Mongols,
Before and after Mars to reign by good luck.

73
The present time together with the past
Will be judged by the great Joker:
The world too late will be tired of him,
And through the clergy oath-taker disloyal.

74
The year of the great seventh number accomplished,
It will appear at the time of the games of slaughter:
Not far from the great millennial age,
When the buried will go out from their tombs.

75
Long awaited he will never return
In Europe, he will appear in Asia:
One of the league issued from the great Hermes,
And he will grow over all the Kings of the East.

76
The great Senate will ordain the triumph
For one who afterwards will be vanquished, driven out:
At the sound of the trumpet of his adherents there will be
Put up for sale their possessions, enemies expelled.

77
Thirty adherents of the order of Quirites
Banished, their possessions given their adversaries:
All their benefits will be taken as misdeeds,
Fleet dispersed, delivered to the Corsairs.

78
Sudden joy to sudden sadness,
It will occur at Rome for the graces embraced:
Grief, cries, tears, weeping, blood, excellent mirth,
Contrary bands surprised and trussed up.

79
The old roads will all be improved,
One will proceed on them to the modern Memphis:
The great Mercury of Hercules fleur-de-lis,
Causing to tremble lands, sea and country.

80
In the realm the great one of the great realm reigning,
Through force of arms the great gates of brass
He will cause to open, the King and Duke joining,
Fort demolished, ship to the bottom, day serene.

81
A treasure placed in a temple by Hesperian citizens,
Therein withdrawn to a secret place:
The hungry bonds to open the temple,

Retaken, ravished, a horrible prey in the midst.

82
Cries, weeping, tears will come with knives,
Seeming to flee, they will deliver a final attack,
Parks around to set up high platforms,
The living pushed back and murdered instantly.

83
The signal to give battle will not be given,
They will be obliged to go out of the park:
The banner around Ghent will be recognized,
Of him who will cause all his followers to be put to death.

84
The illegitimate girl so high, high, not low,
The late return will make the grieved ones contended:
The Reconciled One will not be without debates,
In employing and losing all his time.

85
The old tribune on the point of trembling,
He will be pressed not to deliver the captive:
The will, non-will, speaking the timid evil,
To deliver to his friends lawfully.

86
Like a griffin will come the King of Europe,
Accompanied by those of Aquilon:
He will lead a great troop of red ones and white ones,
And they will go against the King of Babylon.

87
A Great King will come to take port near Nice,
Thus the death of the great empire will be completed:
In Antibes will he place his heifer,
The plunder by sea all will vanish.

88
Foot and Horse at the second watch,
They will make an entry devastating all by sea:
Within the port of Marseilles he will enter,
Tears, cries, and blood, never times so bitter.

89
The walls will be converted from brick to marble,

Seven and fifty pacific years:
Joy to mortals, the aqueduct renewed,
Health, abundance of fruits, joy and mellifluous times.

90
A hundred times will the inhuman tyrant die,
In his place put one learned and mild,
The entire Senate will be under his hand,
He will be vexed by a rash scoundrel.

91
In the year 1609, Roman clergy,
At the beginning of the year you will hold an election:
Of one gray and black issued from Campania,
Never was there one so wicked as he.

92
Before his father the child will be killed,
The father afterwards between ropes of rushes:
The people of Geneva will have exerted themselves,
The chief lying in the middle like a log.

93
The new bark will take trips,
There and near by they will transfer the Empire:
Beaucaire, Arles will retain the hostages,
Near by, two columns of Porphyry found.

94
Scorn from Nîmes, from Arles and Vienne,
Not to obey the Hesperian edict:
To the tormented to condemn the great one,
Six escaped in seraphic garb.

95
To the Spains will come a very powerful King,
By land and sea subjugating the South:
This evil will cause, lowering again the crescent,
Clipping the wings of those of Friday.

96
The Religion of the name of the seas will win out
Against the sect of the son of Adaluncatif:
The stubborn, lamented sect will be afraid
Of the two wounded by A and A.

97

Triremes full of captives of every age,
Good time for bad, the sweet for the bitter:
Prey to the Barbarians hasty they will be too soon,
Anxious to see the feather wail in the wind.

98

For the merry maid the bright splendor
Will shine no longer, for long will she be without salt:
With merchants, bullies, wolves odious,
All confusion universal monster.

99

The end of wolf, lion, ox and ass,
Timid deer they will be with mastiffs:
No longer will the sweet manna fall upon them,
More vigilance and watch for the mastiffs.

100

The great empire will be for England,
The all-powerful one for more than three hundred years:
Great forces to pass by sea and land,
The Lusitanians will not be satisfied thereby.

Almanac: 1555 - 1563

Almanac of 1555

The soul touched from a distance by the divine spirit presages,
Trouble, famine, plague, war to hasten:
Water, droughts, land and sea stained with blood,
Peace, truce, prelates to be born, princes to die.

The Tyrrhenian Sea, the Ocean for the defense,
The great Neptune and his trident soldiers:
Provence secure because of the hand of the great Tende,
More Mars Narbonne the heroic de Villars.

The big bronze one which regulates the time of day,
Upon the death of the Tyrant it will be dismissed:
Tears, laments and cries, waters, ice bread does not give,
V.S.C. peace, the army will pass away.

Near Geneva terror will be great,
Through the counsel, that cannot fail:

The new King has his league prepare,
The young one dies, famine, fear will cause failure.

O cruel Mars, how you should be feared,
More is the scythe with the silver conjoined:
Fleet, forces, water, wind of shadow to fear,
Sea and land in a truce. The friends has joined L.V.

For not having a guard you will be more offended,
The weak fort, Pinquiet uneasy and pacific:
They cry "famine," the people are oppressed,
The sea reddens, the Long one proud and iniquitous.

The five, six, fifteen, late and soon they remain,
The heir's bloodline ended: the cities revolted:
The herald of peace twenty and three return,
The open-hearted five locked up, news invented.

At a distance, near the Aquarius, Saturn turns back,
That year great Mars will give a fire opposition,
Towards the North to the south the great proud female,
Florida in contemplation will hold the port.

Eight, fifteen, and five what disloyalty
The evil spy will come to be permitted:
Fire in the sky, lightning, fear, Papal terror,
The west trembles, pressing too hard the Salty wine.

Six, twelve, thirteen, twenty will speak to the Lady,
The older one by a woman will be corrupted:
Dijon, Guienne hail, lightning makes the first cut into it,
The insatiable one of blood and wine satisfied.

The sky to weep for him, made to do that!
The sea is being prepared, Hannibal to plan his ruse:
Denis [drops anchor], fleet delays, does not remain silent,
Has not known the secret, and by which you are amused!

Venus Neptune will pursue the enterprise,
Pensive one imprisoned, adversaries troubled:
Fleet in the Adriatic, cities towards the Thames,
The fourth clamor, by night, the reposing ones wounded.

The great one of the sky the cape will give,
Relief, Adriatic makes an offer to the port:
He who will be able will save himself from dangers,

By night the Great One wounded pursues.

The port protests too fraudulently and false,
The maw opened, condition of peace:
Rhone in crystal, water, snow, ice stained,
The death, death, wind, through rain the burden broken.

Almanac of 1557

The shocking and infamous armed one will fear the great furnace,
First the chosen one, the captives not returning:
The world's lowest crime, the Angry Female Irale not at east,
Barb, Hister, Malta. And the Empty One does not return.

Conjoined here, in the sky the dispatch opened,
Taken, left behind, mortality not certain:
Little rain, entry, the sky and earth dries,
Undone, death, caught, arrived at a bad hour.

Naval victor at Hoek, Antwerp divorce,
Great heir, from the sky fire, trembling high woods:
Sardinian wood, Malta, Palermo, Corsica,
Prelate to die, strikes the one on the Mule.

The errant herald turns from the dog to the Lion,
Fire will burn the town, pillage, new prize:
To discover foists, Princes taken, they return,
Spy taken Gaul, to the great one joined to the virgin.

From the great court banished, conflict, wounded,
Elected, delivered, accused, cunning mutineers:
And fire on the Pyrenees city, water, venoms, pressed,
Not to travel by water, not to anger the Latins.

Sea, land to go, faith, loyalty broken,
Pillage, wreck, tumult in the city:
Proud, cruel act, ambition sated,
Weakling offended: the perpetrator of the deed unpunished.

Cold, great flood, expelled from the kingdom,
Idiot, discord, Ursa Major and Minor, source in the East:
Poison, siege laid, expelled out of the city,
Happy return, new sect in ruins.

Sea closed, world opened, city exhausted,
The Great One to fail, the newly elected, great mist:
Florence to be open, campus to enter, faith broken,
Stress will be severe to the white plume.

Tutelage on Vesta, war dies, transferred,
Naval combat, honor, death, prelacy:
Death come in, France greatly augmented,
Elected one passed, come to a bad end.

Almanac of 1558

The young King makes a funeral wedding soon,
Holy one stirred up, feasts, of the said, Mars dormant:
Night tears they cry, they conduct the lady outside,
The arrest and peace broken on all sides.

Vain rumor within the Hierarchy,
Genoa to rebel: courses, offenses, tumults:
For the greater King will be the monarchy,
Election, conflict, covert burials.

Through discord in the absence to fail,
One suddenly will put him back on top:
Towards the North will be noises so loud,
Lesions, points to travel, above.

On the Tyrrhenian Sea, of different sail,
On the Ocean there will be diverse assaults:
Plague, poison, blood in the house of canvas,
Prefects, Legates stirred up to march high seas.

There where the faith was it will be broken,
The enemies will feed upon the enemies:
Fire rains [from the] Sky, it will burn, interrupted,
Enterprise by night. Chief will make quarrels.

War, thunder, forces fields, depopulated,
Terror and noise, assault on the frontier:
Great Great One fallen, pardon for the exiles,
Germans, Spaniards, by the sea the Barbarian banner.

The noise will be vain, the faltering ones bundled up,
The Shaven Ones captured: the all-powerful One elected:

The two Reds and four true crusaders to fail,
Rain troublesome to the powerful Monarch.

Rain, wind, forces, Barbarossa Hister, the Tyrrhenian Sea,
Vessels to pass Orkneys and beyond Gibraltar, grain and soldiers provided:
Retreats too well executed by Florence, Siena crossed,
The two will be dead, friendships joined.

Venus the beautiful will enter Florence.
The secret exiles will leave the place behind:
Many widows, they deplore the death of the Great One,
To remove from the realm, the Great Great one does not threaten.

Games, feasts, nuptials, dead Prelate of renown.
Noise, peace of truce while the enemy threatens:
Sea, land and sky noise, deed of the great Brennus,
Cries gold, silver, the enemy they ruin.

Almanac of 1559

Lament, knell, great pillage, to pass the sea, the realm to increase,
Sects, holy ones more polite beyond the sea:
Plague, warmth, fire, banner of the King of Aquilon [the North],
To erect a trophy, city of Henripolis.

The Great One to be no longer, rain, in the chariot, the crystal,
Tumult stirred up, abundance of all goods:
Shaven ones, Holy ones, new ones, old ones frightful,
Elected ingrate, death, lament, joy, alliance.

Grain corrupted, air pestilent, locusts,
Suddenly it will fall, new pasturage to be born:
Captives put in irons, light ones, high-low, loaded,
Through his bones evil which he had not wished to be the King.

Seized in the temple, through a sect's long intrigue,
Elected, ravished in the woods, forms a quarrel:
Seventy pairs new league to be born,
From there their death, King appeased, news.

King hailed as Victor, and Emperor,
The faith broken, the Royal deed known:
Macedonian blood, King made conqueror,

The arrogant people come to humility through tears.

Through spite nuptials, wedding song,
For the three parts Reds, Shaven ones divided:
For the young black/king through fire the soul is restored.
To the great Neptune Ogmios converted.

From the house seven through death in mortal succession,
Hail, tempest, pestilent evil, furies:
King of the East all the West in flight,
He will subjugate his former conquerors.

Pirates pillaged, heat, great drought,
Through too much not being, event not seen, unheard of:
For the foreigner the too great endearment,
New country King, the East fascinated.

The Urn found, the city tributary,
Fields divided, new delusion:
Spain wounded famine, military plaque,
Mockery obstinate, confused, evil, reverie.

Virgins and widows, your good time approaches,
Not at all will it be that which they will pretend:
Far it will be necessary that the approach for it be new,
Very agreeable situation taken, completely restored, it will hold worse.

Here within it will be completed,
The three Great one outside, the Bourbon will be far:
Against the other two one of them will conspire,
At the end of the month they will see the necessity.

Talks held, nuptials recommenced,
The Great Great woman will go out of France:
Voice in Rome not fatigued from crying out,
Receives the peace through too false assurance.

Joy in tears will come to captivate Mars,
Before the Great one the Divines will be stirred up:
Without uttering a word they will enter from three sides,
Mars made drowsy, upon ice run the wines.

Almanac of 1560

Day's journey, diet, interim, no council,
The year peace is being prepared, plague, schismatic famine:
Put outside inside, sky to change, domicile,
End of holiday, hierarchical revolt.

Diet to break up, the ancient sacred one to recover,
Under the two, fire through pardon to result:
Consecration without arms: the tall Red will want to have,
Peace of neglect, the Elected One, the Widower, to live.

To be made to appear elected with novelty,
Place of day-labor to go beyond the boundaries:
The feigned goodness to change to cruelty,
From the suspected place quickly will they all go out.

With the place chosen, the Shaved Ones will not be contented,
Led from Lake Geneva, unproven,
They will cause the old times to be renewed:
They will expose the frighten off the plot so well hatched.

Savoy peace will be broken,
The last hand will cause a strong levy:
The great conspirator will not be corrupted,
And the new alliance approved.

A long comet to wrong the Governor,
Hunger, burning fever, fire and reek of blood:
To all estates Jovial Ones in great honor,
Sedition by the Shaven ones, ignited.

Plague, famine, fire and ardor incessant,
Lightning, great hail, temple struck from the sky:
The Edict, arrest, and grievous law broken,
The chief inventor his people and himself snatched up.

Deprived will be the Shaven Ones of their arms,
It will augment their quarrel much:
Father Liber deceived lightning Albanians,
Sects will be gnawed to the marrow.

The modest request will be received,
They will be driven out and then restored on top:
The Great Great woman will be found content,
Blind ones, deaf ones will be put uppermost.

He will not be placed, the New Ones expelled,

Black king and the Great One will hold hard:
To have recourse to arms. Exiles expelled further,
To sing of victory, not free, consolation.

The mourning left behind, supreme alliances,
Great Shaven One dead, refusal given at the entrance:
Upon return kindness to be in oblivion,
The death of the just one perpetrated at a banquet.

Almanac of 1561

The King, King not to be, destruction by the Clement one,
The year pestilent, the beclouded stirred up:
For the great nobles every man for himself, no joy:
And the term of the mockers will pass.

At the foot of the wall the shy Franciscan,
The enclosure delivered, the cavalry trampling:
Outside the temple Mars and the Scythe of Saturn,
Outside, to divide the friends and upon the reverie.

The times purged, pestilential tempest,
Barbarian insult, fury, invasion:
Infinite evils for this month are prepared for us,
And the Greatest Ones, two less, of mockery.

Joy not long, abandoned by his followers,
The year pestilent, the Greatest One assailed:
The good Lady in the Elysian Fields,
And the greater part of the cold goods not gathered.

Incursions of the Lion, not to prepare for conflicts,
Sad enterprise, the air pestilent, hideous:
From all sides the Great Ones will be afflicted,
And ten and seven to assail twenty and two.

Retaken, surrendered, terror-stricken by the evil,
The blood far inferior, and the faces hideous:
To the most knowledgeable ones the ignorant one frightful,
Plague, hatred, horror, the piteous female will fall low.

Dead and seized, the nonchalant ones of the exchange,
They will go far away in approaching more strongly:

United ones locked up in the ruin, barn,
Through long help the strongest one astonished.

Gray, whites and blacks, hidden, and broken,
They will be put off, divided, put in their sieges:
The ravishers will find themselves mocked,
And the Vestals confined behind strong bars.

Almanac of 1562

Season of winter, good spring, sound, bad summer,
Pernicious autumn, dry, wheat rare:
Of wine enough, bad eyes, deeds, molested,
War, mutiny, seditious waste.

The hidden desire for the good will succeed,
Religion, peace, love and concord:
The nuptial song will not be completely in accord,
The high ones, who are low, and high, put to the rope.

For the Shaven Ones the Chief will not reach the end,
Edicts changed, the secret ones set at large:
Great One found dead, less of faith, low standing,
Dissimulated, shuddering, wounded in the boar's lair.

Moved by Lion, near Lion he will undermine,
Taken, captive, pacified by a woman:
He will not hold as well as they will waver,
Placed unpassed, to oust the soul from rage.

From Lion he will come to arouse to move,
Vain discovery against infinite people:
Known by none the evil for the duty,
In the kitchen found dead and finished.

Nothing in accord, worse and more severe trouble,
As it was, land and sea to quieten:
All arrested, it will not be worth a double,
The iniquitous one will speak, Counsel of annihilation.

Portentous deed, horrible and incredible,
Typhoon will make the wicked ones move:
Those who then afterwards supported by the cable,
And the greater part exiled on the fields.

Right put on the throne come into France from the sky,
The whole world pacified by Virtue:
Much blood to scatter, sooner change to come,
By the birds, and by fire, and not by vers.

The colored ones, the Sacred malcontents,
Then suddenly through the happy Androgynes:
Of the great part to see, the time not come,
Several amongst them will make their soups weak.

They will be returned to their full power,
Conjoined at one point of the accord, not in accord:
All defied, more promised to the Shaven Ones,
Several amongst them outflanked in a band.

For the legate of terrestrial and dawn,
The great Cape will accommodate himself to all:
Tacit LORRAINE, to be listening,
He whose advice they will not want to agree with.

The enemy wind will impede the troop,
For the greatest one advance put in difficulty:
Wine with poison will be put in the cup,
To pass the great gun without horse-power.

Through crystal the enterprise is broken,
Games and feats, in LYONS to repose more:
No longer will he take his repast with the Great Ones,
Sudden catarrh, blessed water, to bathe him.

Almanac: 1564 – 1567

Almanac of 1564

The sextile year rains, wheat to abound, hatreds,
Joy to men, Princes, King divorced:
Herd to perish, human mutations,
People oppressed and poison under the surface.

Times very diverse, discord discovered,
Council of war, change taken in, changed:
The Great Woman must not be, conspirators through water lost,
Great hostility, for the great one all steady.

The bit of the enemy's tongue approaches,
The Debonair one to peace will want to reduce:
The obstinate ones will want to lose the kinswoman,
Surprised, Captives, and suspects fury to injure.

Fathers and mothers dead of infinite sorrows,
Women in mourning, the pestilent she-monster:
The Great One to be no more, all the world to end,
Under peace, repose and every single one in opposition.

Princes and Christendom stirred up in debates,
Foreign nobles, Christ's See molested:
Become very evil, much good, mortal sight.
Death in the East, plague, famine, evil treaty.

Land to tremble, killed, wasteful, monster,
Captives without number, to do, undone, done:
To go over the sea misfortune will occur,
Proud against the proud evil done in disguise.

The unjust one lowered, they will molest him fiercely,
Hail, to flood, treasure, and engraved marble:
Chief of Persuasion people will kill to death,
And attached will be the blade to the tree.

Of what not evil? inexcusable result,
The fire not double, the Legate outside confused:
Against the worse wounded the fight will not be made,
The end of June the thread cut by firing.

Fine bonds enfeebled by accords,
Mars and Prelates united will not stop:
The great ones confused by gifts of mutilated bodies,
Dignified ones, undignified ones will seize the well endowed.

From good to the evil times will change,
The peace in the South, the expectation of the Greatest Ones:
The Great Ones grieving Louis too much more will stumble,
Well-known Shaven Ones have neither power not understanding.

This is the month for evils so many as to be doubled,
Deaths, plague to drain all, famine, to quarrel:
Those of the reverse of exile will come to note,
Great Ones, secrets, deaths, not to censure.

Through death, death to bite, counsel, robbery, pestiferous,

They will not dare to attack the Marines:
Deucalion a final trouble to make,
Few young people: half-dead to give a start.

Dead through spite he will cause the others to shine,
And in an exalted place some great evils to occur:
Sad concepts will come to harm each one,
Temporal dignified, the Mass to succeed.

Almanac of 1565

A hundred times worse this year than the year passed,
Even for the Greatest Ones of the realm and of the Church:
Infinite evils, death, exile, ruin, smashed,
To death great woman to be, pestilence, plague sores, and bile.

Snow, rustiness, rains and great rains,
Even for the Greatest One joy, pestilence, sleepless:
Seeds, plentiful grains, and more the bands,
They will prepare themselves, enmity unallayed.

Between the Great Ones a great discord will arise,
The noble Cleric will plot a great event:
New sects to place in hatred and discord,
All people will strive for war and change.

Secret conspiracy, perilous change,
Factions to conspire secretly:
Rains, great winds, for arrogant ones,
Rivers to overflow, pernicious actions.

Plague to multiply, the Sects to quarrel with each other,
Times moderated, winter, little return:
Of mass and meeting house grievously to debate,
Rivers to overflow, evils, deadly all around.

To the people of lower class in debates and quarrels,
And through the women and defunct ones great war:
Death of a Great Woman, to celebrate the scrofula,
More great Ladies expelled from the land.

In widowhood as many males as females,
The life of great monarchs to be in jeopardy:
Plague, steel, famine, great peril, pell-mell,

Troubles through changes, lesser nobility to incite.

Hail, rust, rains, and great plagues,
To preserve women, they will be the cause of the noise:
Death of several through plague, steel, famine, through hatreds,
The heavens will be seen, which is to say that it will be relit.

Not at all sufficient will be the grain,
The death approaches to snow more than white:
Sterility, grain rotted, the water abundant,
The great one wounded, several put to death on the flank.

War, of fruits nor grain, trees and scrub-brush,
Great volatility, noble to spur on:
As temporal as the young-lion-like prelate,
D'ANDELOT to conquer, noble ones to draw back.

Everything changed, one to persecute four,
Outside malady, morality very far:
Of the four two more will not come to debate,
Exile, ruin, death, famine, perplexity.

Of the great ones the greater number will not be so many,
Great changes, commotions, steel, plague:
The small estimate: lent, paid, content,
Opposite month frost molests severely.

Severe frost, ice more than concord,
Widows, matrons, fire, lament:
Games, frolic, joy, Mars to incite discord,
Through marriage good expectations.

Almanac of 1566

For the greatest ones death, loss of honor and violence,
Professors of the faith, their estate and their sect:
For the two great Churches diverse noise, decadence,
Evil neighbors quarreling serfs of the Church without a head.

Waste, great loss, and not without violence,
All those of the faith, more for religion,
The Greatest Ones will lose their lives, their honor and fortunes
Both the two Churches, the sin in their faction.

For the two very Great Ones pernicious loss to arise,
The Greatest Ones will cause loss, goods, of honor, and of life,
As much great noises will run, the urn very odious,
Great maladies to be, meeting-house, mass in envy.

The servants of the Churches will betray their Lords,
Of other Lords also by the undivided of the fields:
Neighbors of meeting-house and mass will quarrel amongst them,
Rumors, noises to augment, to death are several lying.

Of all blessings abundance, the earth will produce for us,
No din of war in France, sedition put outside:
Man-slayers, robbers one will find on the highway,
Little faith, burning fever, people in commotion.

Between people discord, brutal enmity,
War, death of great Princes, several parts:
Universal plague, stronger in the West,
Times good and full, but very dry and exhausted.

The grains not to be plentiful, in all other fruits, plenty,
The Summer, spring humid, winter long, snow, ice:
The East in arms, France reinforces herself,
Death of beasts much honey, the place to be besieged.

Through pestilence and fire fruits of trees will perish,
Signs of oil to abound. Father Denis not scarce:
Some great ones to die, but few foreigners will sally forth in attack,
Offense, Barbarian marines, and dangers at the frontiers.

Rains very excessive, and of blessings abundance,
The cattle price to be just, women outside of danger:
Hail, rain, thunder: people depressed in France,
Through death they will work, death to reprove people.

Arms, plagues to cease, death of the seditious ones,
Great Father Liber will not much abound:
Evil ones will be seized by more malicious ones,
France more than ever victorious will triumph.

Up to this month the great drought will endure,
For Italy and Provence all fruits to half:
The Great One less of enemies prisoner of their band,
For the scroungers, Pirates, and the enemy to die.

The enemy so much to be feared to retire into Thrace,

Leaving cries, howls, and pillage desolated:
To leave noise on sea and land, religion murdered,
Jovial Ones put on the road, every sect to become angry.

Almanac of 1567

Death, malady for young women, head colds,
From the head to the eyes a wretched deal of land:
By sea misfortune, seeds bad, wine in mists,
Much oil too much rain, molests the fruits, war.

Prisons, secrets: annoyances, discord between neighbors,
They will give life, through evil diverse catarrhs:
Death will ensue, poison will cause concord,
Terror, fear, great dread traveling will release from guarantees.

Prisons for enemies hidden and manifested,
Travel will not hold, mortal enmity:
Love three, secret hostility, public festival,
The broken ruined, the water will break the quarrel.

The public enemies, nuptials and marriages,
Death after, he grown rich through the deaths:
The great friends will show themselves in the passage,
Two sects to jabber, from surprise remorse later.

Through great maladies, religion offended,
Through the infants and gifts of the Embassy:
Gift given to worthless one, new law relaxed,
Goods of old fathers, King in good country.

From the father it approaches the son: Magistrates called severe,
The great nuptials, enemies mangling:
The concealed put in front, through the faith of reproaches,
The good friends and women against such grumblings.

Through the treasure, found -- the heritage of the father,
The Kings and Magistrates, the nuptials, enemies:
The public malevolent, the Judge and Mayor,
The death, fear and terror: and three Great Ones put to death.

Again the death approaches, Royal gift and Legacy,
He will prepare what is, through old age in decay:
The young heirs in suspicion of no legacy,

Treasure found in plasters and kitchen cookery.

The secret enemies will be imprisoned,
The Kings and Magistrates will hold there a sure hand:
The life of several, healthiness, malady in eyes and nose,
The two great ones will go away very far at the bad hour.

Long debility in the head, nuptial, enemy,
Through Prelate and journey, dream of the Great One terror:
Fire and ruin, great one found in the oblique place,
By torrent discovered, new error to come out.

The Kings and Magistrates through the deaths, the hand to place,
Young girls sick, and of the Great Ones the body swells:
All through languors and nuptials, enemy serfs for the master,
The public sad, the Composer all swelled up.

On his return from the Embassy, the King's gift put in place.
He will do nothing more. He will be gone to God.
Close relatives, friends, brothers by blood,
[Will find him] Completely dead near the bed and the bench.

The Life of Nostradomus

A Short Biography

Michel de Nostradame was born on December 14, 1503 at St. Remy in Provence, France to a family of   He was taught a wide range of subjects by both his grandfathers. By the time that Michel began his formal education in Avignon, where he learned philosophy, grammar, and rhetoric, he was already well versed in classical literature, history, medicine, astrology (then a legitimate science), and herbal and folk medicine.
Nostradamus (the Latin version of his name) first became well known due to his novel and very successful treatment of bubonic plague, the infamous *Black Death* that ravaged France in the early 16th century. His innovative cure consisted basically of cleanliness and vitamin C. The first step when he entered a village was to have all of the corpses removed from the streets. He then prescribed for his patients plenty of fresh air, unpolluted water, and his "Rose Pills," which consisted of rose petals (and, perhaps, rose hips), sawdust from green cypress, iris, cloves, calamus, and lign-aloes. He did not allow his patients to suffer "bleeding," then a popular, although futile, treatment for everything from a minor cold to the Great Plague. He truly was successful in combating a disease for which there appeared to be no cure and no relief. It is estimated that over one-quarter of the entire population of Europe was killed by the Black Death during its sporadic visits.
In 1537, plague struck Agen, where Nostradamus was living with his wife and two children. He confidently began to treat his fellow citizens but, unfortunately, was unable

to save his family from his old enemy. Distraught and questioning his own abilities, Nostradamus began to wander through Europe aimlessly for the next six years. It was during this time that he first became aware of the awakening of his prophetic powers. When plague broke out in Aix, capital of Provence, for nine months Nostradamus again applied his proven skills to save as many of the populace as possible. After the plague dissipated, the city showed its gratitude by bestowing upon him a pension for life.

Ten years after the death of his wife and children, Nostradamus settled in Salon and remarried, eventually fathering three daughters and three sons. He had the upper floor of his house transformed into a private study, where he installed his magical equipment: astrolabe, magic mirrors, divining rods, and a brass bowl and tripod, designed after the type used by the great Delphic oracles.

Under cover of night, he would retire to his study where he would sit before the tripod upon which simmered the brass bowl filled with water and pungent herbs (probably bay laurel or a combination of herbs). For a few years, Nostradamus struggled with the dilemma of whether he should make his findings public. In 1550, he published his first almanac of prophecies -- twelve four-line poems called "quatrains." Each quatrain gave a general prophecy for the coming year. The acclaim that he received due to the almanac encouraged Nostradamus to continue. He produced an almanac every year for the rest of his life.

His most famous work, *The Centuries*, was begun in 1554. Eventually these prophecies were to consist of ten volumes of 100 quatrains each. Centuries 1 through part of 4 were published in Lyons in 1555. The remainder of 4 and the subsequent Centuries through 7 were published later that same year. The last three were printed in 1558, but Nostradamus decided not to have them widely distributed. *The Centuries* have remained constantly in print for over 400 years.

In his own time as is true today, Nostradamus' quatrains have received mixed responses. The combination of French, Provencal, Greek, Latin, and Italian written as riddles, puns, anagrams, and epigrams are complex and demand that the potential interpreter have knowledge in a wide range of subjects. Many people from the 16th century through modern times have been enthralled by the prophecies and have tried to make sense of them. Some quatrains could fit descriptions of just about any era. Others are more exact, and it is those quatrains that have established the well-earned reputation of Nostradamus as one of the world's greatest prophets.

To the ignorant, closed mind, Nostradamus was a creature of the devil, babbling in cryptic, evil verse. From philosophers, Nostradamus continues to draw praise and curses. Poets remain perplexed by the meaning of his wild verses. Interpretation is open to all. Except for those prophecies that were fulfilled in his own time and acknowledged by the prophet himself, no one can give a final, definite interpretation to any of the prophecies that have yet to be fulfilled. This is an area that is open to individual, solitary study. Perhaps that was the intent.

According to witnesses, Nostradamus stayed alert to the end of his life, even though he was in great pain caused by arthritis, gout and dropsy. When his assistant wished him goodnight on July 1, 1566, Nostradamus replied, "You will not find me alive at sunrise." As one might expect, he had predicted his own death. In his last almanac, Nostradamus had written:

On his return from the Embassy, the King's gift put in place,
He will do nothing more. He will be gone to God.
Close relatives, friends, brothers by blood (will find him)
Completely dead near the bed and the bench.
On the morning of July 2, the assistant escorted family and friends to the study, where Nostradamus had spent the previous night. They found his body on the floor between the bed and a bench that he had placed there for aid in getting out of bed.
His wife carried out his last wishes, that he be interred standing upright and that his coffin be enclosed within the walls of the Church of the Cordeliers of Salon. The translation from Latin of the inscription on his tomb reads:
"Here rest the bones of the illustrious Michel Nostradamus, alone of all mortals, judged worthy to record with his near divine pen, under the influence of the stars, the future events of the entire world. He lived sixty-two years, six months and seventeen days. He died at Salon the in year 1566. Let not posterity disturb his rest. Anne Posart Gemelle wishes her husband true happiness."
[From: by John Hogue]

His Method

Nostradamus used a variety of magical arts and tools to induce ecstatic trances. Visions came to him through flame or water gazing, sometimes both together. he also followed the practice of Branchus, a   priestess of ancient Greece, requiring him to sit, spine erect, on a brass tripod, the legs of which were angled at the same degree as the Egyptian pyramids. The upright position, and possibly the use of nutmeg (a mild hallucinogen when consumed in sufficient quantity -- [*deadly when the dose is too large*]), stimulated the mind; the angle of the tripod legs was thought to create a bioelectric force which would sharpen psychic powers.
Or the prophet would stand or sit before a tripod that held a brass bowl filled with steaming water and pungent oils. "I emptied my soul, brain and heart of all care and attained a state of tranquillity and stillness of mind which are prerequisites for predicting by means of the brass tripod."
"Although the everlasting God alone knows the eternity of light proceeding from himself, I say frankly to all to whom he wishes to reveal his immense magnitude -- infinite and unknowable as it is -- after long and meditative inspiration, that it is a hidden thing divinely manifested to the prophet by two means: One comes by infusion which clarifies the supernatural light in the one who predicts by the stars, making possible *divine revelation*; the other comes by means of participation with the divine eternity, by which means the prophet can judge what is given from his (her) *own divine spirit* through God the Creator and *natural intuition*."
[Preceding from   p. 34, by John Hogue. Above italics are mine.]
My own independently arrived at conclusion is that no magic is required for basic prophecy; prophetic ability isn't very eerie at all. Intense meditation would, of course, have been necessary for a prophet on the order of Nostradamus, one who was looking into the very far future. Ritual tools might have helped his concentration.

The art of prophecy seems (to me) to be comprised of studied, practiced mastery of the following (not necessarily in order of importance):
A thorough knowledge of the cycles of "history";
Careful, detached observation of the recent past and the present;
More than passing familiarity with basic human nature;
Meditation, contemplation -- ability to eliminate one's own thoughts, desires, biases;
Inspiration;
Intuition, common sense, and compassion.
Keep in mind that a fulfilled prophecy is a failed prophecy. The great prophets didn't broadcast their predictions so that they could sit back and laugh, "I told you so!" They were issuing warnings about visions that, for the most part, they would rather not have seen.
Personally, I don't trust the results of many of today's so-called *prophets* or *experts* who charge high fees for seminar attendance, issue press releases, and mail or hand out brochures boasting about their own "success rates."
While I'm on the subject, those of you who are familiar with Dolores Cannon, a present-day Nostradamus "interpreter," might notice that I haven't used her work here. This is because her interpretations are purportedly "channeled" material. I do not discount all channeling. However, there's too much room for error. The channel might be innocently "channeling" her/his own subconscious thoughts and desires. More to the point, outright fraud is inherent in the field, unfortunately. If Ms. Cannon had published her interpretations as educated guesses, I would have given them more than a casual glance. This is not to say that I am accusing her of any malevolence.

Anecdotal Stories

A group of Franciscan monks traveling one day along a muddy road near the Italian town of Ancona suddenly saw the solitary doctor walking toward them. Nostradamus stepped aside to let them pass but, before they had done so, knelt in the mud in front of one of them, Brother Felice Peretti. The friars were puzzled by this. Peretti was of lowly birth and had been a swineherd before joining the Order of Saint Francis. Nostradamus told them, "I must cede myself and bend a knee before his Holiness."
The monks reacted with understandable amusement. Nostradamus must have appeared to be a madman. This probably occurred during the six years that he wandered through Europe after the Black Death had taken his family. However, forty years after this chance meeting on the muddy road and nineteen years after the death of Nostradamus, Brother Peretti was elected Pope Sixtus V.

While Nostradamus was visiting the chateau of Lord de Florinville, he had conversations with his host about prophecy. Florinville decided to put the prophet to a test. At the time, they had stopped during their stroll and stood before a corral enclosing two suckling pigs -- one black and one white. The host asked Nostradamus which pig would be served for dinner that night. "We will eat the black pig, but a wolf will eat the white," was the reply.

Florinville secretly ordered his cook to prepare the white pig for dinner that night. The cook followed the orders but left the door to the kitchen open while running out on another errand. When he returned, he found the chateau's pet wolf eating the already dressed white pig. Worried about the results of his error, the cook quietly prepared the black pig for the night's meal.

At the dinner table, Lord de Florinville smiled broadly at Nostradamus and announced, "We are not eating the black pig as you predicted. And no wolf will touch our dinner here."

Nostradamus was so sure that this was the black pig that his host summoned the cook to prove him wrong. Of course, Florinville was stunned when the cook delivered the bad news.

Century 1, Quatrain 65
The young lion will overcome the older one
On the field of combat in single battle
He will pierce his eyes through a golden cage
Two wounds made one, then he dies a cruel death.

This prophecy was already well known in 1559 when Henry II of France held a three-day knightly tournament in honor of the marriages of his sister Marguerite to the Duke of Savoy and of his daughter to King Philip II of Spain. Henry participated in the events, resplendent in full armor, carrying his great shield decorated with an ornate lion. After winning each round, he would raise the visor of his golden helmet to receive the praises of the crowd.

On the third day, at sunset, Henry prepared for his final bout against Count Montgomery. The bout ended in a draw and, when Henry insisted on a final match, the young count tried to excuse himself, aware of the prophecy. After Henry continued to insist, Montgomery relented.

During the second charge, there was loud crack of broken lances. A splinter from the count's broken lance pierced the king's golden visor and lodged behind his left eye, blinding him and penetrating deep into his brain. He lingered for ten days in agony before dying and fulfilling one of Nostradamus' most famous prophecies.

Desecration of the Tomb of Nostradomus

Century 9, Quatrain 7
The man who opens the tomb when it is found
And who does not close it immediately,
Evil will come to him
That no one will be able to prove.

Anne Gemelle, Nostradamus' wife, carried out his last wishes concerning the disposal of his body. He was entombed upright in a wall of the Church of the Cordeliers in Salon, France to ensure that his detractors would not be able to "put your filthy feet on my throat while I'm alive or after I'm dead."

His resting place became a pilgrimage site soon after his entombment, and for centuries, a rumor circulated that the prophet had had a secret document, giving the keys to deciphering the quatrains, buried with him. In 1700, city officials decided to move his body to an area behind a more prominent wall of the church. While they had the tomb open, they couldn't resist a careful peek inside the coffin.

No papers were found, unfortunately. However, a medallion hung around the skeleton's neck. This medallion had been inscribed with the date 1700. Nostradamus had a last laugh by predicting, in 1566, when his tomb would be invaded. The tomb was resealed and undisturbed for another 91 years.

In 1791, during the French Revolution, drunken soldiers broke into the church and, using picks and shovels, looted the tomb of Nostradamus. The sound of the commotion alerted the mayor of Salon, who hurried to the church to investigate. He arrived to witness the ghastly scene of soldiers and townspeople tossing the prophet's bones into the air, dancing in macabre drunkenness. One guardsman stood in the center of the group, drinking wine from the skull of Nostradamus. (Local people believed that drinking blood from the skull of the great prophet would bestow psychic abilities.) Nostradamus had warned that anyone who dared disturb his rest would suffer a quick and violent death. The mayor acted quickly and explained to the soldiers that Nostradamus, in having predicted the French Revolution in supportive tones, should be considered a national hero. Those present collected the bones from the floor and helped to reinter the remains. According to the legend, these revolutionary soldiers were ambushed by royalists while returning to their base in Marseilles. The soldier who had brazenly drunk from Nostradamus' skull was killed, quickly and violently, by a sniper's bullet.

In Paris, for ten days following the storming of the Bastille on July 14, 1789, visitors to the fortress filed past a table upon which was a copy of *The Centuries* opened to the page of Nostradamus' predictions describing the French Revolution ("Common Advent of the People"), written over 200 years earlier.

This is an accurate translation of the Preface by M. Nostradamus to his Prophecies. He wrote it as a letter to his son Cesar.
[Found at: http://www.astrologer.ru:8003/Nostradamiana/centuries-eng.html, which also provides the French version.]

Preface by M. Nostradamus to His Prophecies

Greetings and happiness to Cesar Nostradamus my son. Your late arrival, Cesar Nostredame, my son, has made me spend much time in constant nightly reflection so that I could communicate with you by letter and leave you this reminder, after my death, for the benefit of all men, of which the divine spirit has vouchsafed me to know by means of astronomy. And since it was the Almighty's will that you were not born here in this region and I do not want to talk of years to come but of the months during which you will struggle to grasp and understand the work I shall be compelled to leave you after my death: assuming that it will not be possible for me to leave you such writing as may be destroyed through the injustice of the age. The key to the hidden prediction which you will inherit will be locked inside my heart.

Also bear in mind that the events here described have not yet come to pass, and that all is ruled and governed by the power of Almighty God, inspiring us not by bacchic frenzy nor by enchantments but by astronomical assurances: predictions have been made through the inspiration of divine will alone and the spirit of prophecy in particular.

On numerous occasions and over a long period of time I have predicted specific events far in advance, attributing all to the workings of divine power and inspiration, together with other fortunate or unfortunate happenings, foreseen in their full unexpectedness, which have already come to pass in various regions of the earth. Yet I have wished to remain silent and abandon my work because of the injustice not only of the present time but also for most of the future. I will not commit to writing.

Since governments, sects and countries will undergo such sweeping changes, diametrically opposed to what now obtains, that were I to relate events to come, those in power now - monarchs, leaders of sects and religions - would find these so different from their own imaginings that they would be led to condemn what later centuries will learn how to see and understand. Bear in mind also Our Saviour's words: "Do not give anything holy to the dogs, nor throw pearls in front of swine lest they trample them with their feet and turn on you and tear you apart." For this reason I withdrew my pen from the paper, because I wished to amplify my statement touching the Vulgar Advent by means of ambiguous and enigmatic comments about future causes, even those closest to us and those I have perceived, so that some human change which may

Writings of Nostradamus

Nostradamus

Table of Contents

Writings of Nostradamus..1
    Nostradamus..1
    Preface..1
    Century I..5
    Century II..17
    Century III..28
    Century IV..40
    Century V..52
    Century VI..64
    Epistle to Henry II..76
    Century VII..83
    Century VIII..88
    Century IX..99
    Century X..111
    Almanacs: 1555–1563..123
    Almanacs: 1564–1567..128

Writings of Nostradamus

Nostradamus

- Preface
- Century I
- Century II
- Century III
- Century IV
- Century V
- Century VI
- Epistle to Henry II
- Century VII
- Century VIII
- Century IX
- Century X
- Almanacs: 1555–1563
- Almanacs: 1564–1567

Preface

Preface by M. Nostradamus to His Prophecies

Greetings and happiness to Cesar Nostradamus my son. Your late arrival, Cesar Nostredame, my son, has made me spend much time in constant nightly reflection so that I could communicate with you by letter and leave you this reminder, after my death, for the benefit of all men, of which the divine spirit has vouchsafed me to know by means of astronomy. And since it was the Almighty's will that you were not born here in this region and I do not want to talk of years to come but of the months during which you will struggle to grasp and understand the work I shall be compelled to leave you after my death: assuming that it will not be possible for me to leave you such writing as may be destroyed through the injustice of the age. The key to the hidden prediction which you will inherit will be locked inside my heart.

Also bear in mind that the events here described have not yet come to pass, and that all is ruled and governed by the power of Almighty God, inspiring us not by bacchic frenzy nor by enchantments but by astronomical assurances: predictions have been made through the inspiration of divine will alone and the spirit of prophecy in particular.

On numerous occasions and over a long period of time I have predicted specific events far in advance, attributing all to the workings of divine power and inspiration, together with other fortunate or unfortunate happenings, foreseen in their full unexpectedness, which have already come to pass in various regions of the earth. Yet I have wished to remain silent and abandon my work because of the injustice not only of the present time but also for most of the future. I will not commit to writing.

Writings of Nostradamus

Since governments, sects and countries will undergo such sweeping changes, diametrically opposed to what now obtains, that were I to relate events to come, those in power now – monarchs, leaders of sects and religions – would find these so different from their own imaginings that they would be led to condemn what later centuries will learn how to see and understand. Bear in mind also Our Saviour's words: "Do not give anything holy to the dogs, nor throw pearls in front of swine lest they trample them with their feet and turn on you and tear you apart." For this reason I withdrew my pen from the paper, because I wished to amplify my statement touching the Vulgar Advent by means of ambiguous and enigmatic comments about future causes, even those closest to us and those I have perceived, so that some human change which may come to pass shall not unduly scandalize delicate sensibilities. The whole work is thus written in a nebulous rather than plainly prophetic form. So much so that, "You have hidden these things from the wise and the circumspect, that is from the mighty and the rulers, and you have purified those things for the small and the poor," and through Almighty God's will, revealed unto those prophets with the power to perceive what is distant and thereby to foretell things to come. For nothing can be accomplished without this faculty, whose power and goodness work so strongly in those to whom it is given that, while they contemplate within themselves, these powers are subject to other influences arising from the force of good. This warmth and strength of prophecy invests us with its influence as the sun's rays affect both animate and inanimate entities.

We human beings cannot through our natural consciousness and intelligence know anything of God the Creator's hidden secrets, For it is not for us to know the times or the instants, etc.

So much so that persons of future times may be seen in present ones, because God Almighty has wished to reveal them by means of images, together with various secrets of the future vouchsafed to orthodox astrology, as was the case in the past, so that a measure of power and divination passed through them, the flame of the spirit inspiring them to pronounce upon inspiration both human and divine. God may bring into being divine works, which are absolute; there is another level, that of angelic works; and a third way, that of the evildoers.

But my son, I address you here a little too obscurely. As regards the occult prophecies one is vouchsafed through the subtle spirit of fire, which the understanding sometimes stirs through contemplation of the distant stars as if in vigil, likewise by means of pronouncements, one finds oneself surprised at producing writings without fear of being stricken for such impudent loquacity. The reason is that all this proceeds from the divine power of Almighty God from whom all bounty proceeds.

And so once again, my son, if I have eschewed the word prophet, I do not wish to attribute to myself such lofty title at the present time, for whoever is called a prophet now was once called a seer; since a prophet, my son, is properly speaking one who sees distant things through a natural knowledge of all creatures. And it can happen that the prophet bringing about the perfect light of prophecy may make manifest things both human and divine, because this cannot be done otherwise, given that the effects of predicting the future extend far off into time.

God's mysteries are incomprehensible and the power to influence events is bound up with the great expanse of natural knowledge, having its nearest most immediate origin in free will and describing future events which cannot be understood simply through being revealed. Neither can they be grasped through men's interpretations nor through another mode of cognizance or occult power under the firmament, neither in the present nor in the total eternity to come But bringing about such an indivisible eternity through Herculean efforts, things are revealed by the planetary movements.

I am not saying, my son – mark me well, here – that knowledge of such things cannot be implanted in your deficient mind, or that events in the distant future may not be within the understanding of any reasoning being. Nevertheless, if these things current or distant are brought to the awareness of this reasoning and intelligent being they will be neither too obscure nor too clearly revealed.

Writings of Nostradamus

Perfect knowledge of such things cannot be acquired without divine inspiration, given that all prophetic inspiration derives its initial origin from God Almighty, then from chance and nature. Since all these portents are produced impartially, prophecy comes to pass partly as predicted. For understanding created by the intellect cannot be acquired by means of the occult, only by the aid of the zodiac, bringing forth that small flame by whose light part of the future may be discerned.

Also, my son, I beseech you not to exercise your mind upon such reveries and vanities as drain the body and incur the soul's perdition, and which trouble our feeble frames. Above all avoid the vanity of that most execrable magic formerly reproved by the Holy Scriptures – only excepting the use of official astrology.

For by the latter, with the help of inspiration and divine revelation, and continual calculations, I have set down my prophecies in writing. Fearing lest this occult philosophy be condemned, I did not therefore wish to make known its dire import; also fearful that several books which had lain hidden for long centuries might be discovered, and of what might become of them, after reading them I presented them to Vulcan. And while he devoured them, the flame licking the air gave out such an unexpected light, clearer than that of an ordinary flame and resembling fire from some flashing cataclysm, and suddenly illumined the house as if it were caught in a furnace. Which is why I reduced them to ashes then, so that none might be tempted to use occult labours in searching for the perfect transmutation, whether lunar or solar, of incorruptible metals.

But as to that discernment which can be achieved by the aid of planetary scrutiny, I should like to tell you this. Eschewing any fantastic imaginings, you may through good judgement have insight into the future if you keep to the specific names of places that accord with planetary configurations, and with inspiration places and aspects yield up hidden properties, namely that power in whose presence the three times are understood as Eternity whose unfolding contains them all: for all things are naked and open.

That is why, my son, you can easily, despite your young brain, understand that events can be foretold naturally by the heavenly bodies and by the spirit of prophecy: I do not wish to ascribe to myself the title and role of prophet, but emphasize inspiration revealed to a mortal man whose perception is no further from heaven than the feet are from the earth. I cannot fail, err or be deceived, although I may be as great a sinner as anyone else upon this earth and subject to all human afflictions.

But after being surprised sometimes by day while in a trance, and having long fallen into the habit of agreeable nocturnal studies, I have composed books of prophecies, each containing one hundred astronomical quatrains, which I want to condense somewhat obscurely. The work comprises prophecies from today to the year 3797.

This may perturb some, when they see such a long timespan, and this will occur and be understood in all the fullness of the Republic; these things will be universally understood upon earth, my son. If you live the normal lifetime of man you will know upon your own soil, under your native sky, how future events are to turn out.

For only Eternal God knows the eternity of His light which proceeds from Him, and I speak frankly to those to whom His immeasurable, immense and incomprehensible greatness has been disposed to grant revelations through long, melancholy inspiration, that with the aid of this hidden element manifested by God, there are two principal factors which make up the prophet's intelligence.

The first is when the supernatural light fills and illuminates the person who predicts by astral science, while the second allows him to prophesy through inspired revelation, which is only a part of the divine eternity, whereby the prophet comes to assess what his divinatory power has given him through the grace of God and by a natural gift, namely, that what is foretold is true and ethereal in origin.

Writings of Nostradamus

And such a light and small flame is of great efficacy and scope, and nothing less than the clarity of nature itself. The light of human nature makes the philosophers so sure of themselves that with the principles of the first cause they reach the loftiest doctrines and the deepest abysses.

But my son, lest I venture too far for your future perception, be aware that men of letters shall make grand and usually boastful claims about the way I interpreted the world, before the worldwide conflagration which is to bring so many catastrophes and such revolutions that scarcely any lands will not be covered by water, and this will last until all has perished save history and geography themselves. This is why, before and after these revolutions in various countries, the rains will be so diminished and such abundance of fire and fiery missiles shall fall from the heavens that nothing shall escape the holocaust. And this will occur before the last conflagration.

For before war ends the century and in its final stages it will hold the century under its sway. Some countries will be in the grip of revolution for several years, and others ruined for a still longer period. And now that we are in a republican era, with Almighty God's aid, and before completing its full cycle, the monarchy will return, then the Golden Age. For according to the celestial signs, the Golden Age shall return, and after all calculations, with the world near to an all-encompassing revolution – from the time of writing 177 years 3 months 11 days – plague, long famine and wars, and still more floods from now until the stated time. Before and after these, humanity shall several times be so severely diminished that scarcely anyone shall be found who wishes to take over the fields, which shall become free where they had previously been tied.

This will be after the visible judgement of heaven, before we reach the millennium which shall complete all. In the firmament of the eighth sphere, a dimension whereon Almighty God will complete the revolution, and where the constellations will resume their motion which will render the earth stable and firm, but only if He will remain unchanged for ever until His will be done.

This is in spite of all the ambiguous opinions surpassing all natural reason, expressed by Mahomet; which is why God the Creator, through the ministry of his fiery agents with their flames, will come to propose to our perceptions as well as our eyes the reasons for future predictions.

Signs of events to come must be manifested to whomever prophesies. For prophecy which stems from exterior illumination is part of that light and seeks to ally with it and bring it into being so that the part which seems to possess the faculty of understanding is not subject to a sickness of the mind.

Reason is only too evident. Everything is predicted by divine afflatus and thanks to an angelic spirit inspiring the one prophesying, consecrating his predictions through divine unction. It also divests him of all fantasies by means of various nocturnal apparitions, while with daily certainty he prophesies through the science of astronomy, with the aid of sacred prophecy, his only consideration being his courage in freedom.

So come, my son, strive to understand what I have found out through my calculations which accord with revealed inspiration, because now the sword of death approaches us, with pestilence and war more horrible than there has ever been – because of three men's work – and famine. And this sword shall smite the earth and return to it often, for the stars confirm this upheaval and it is also written: "I shall punish their injustices with iron rods, and shall strike them with blows."

For God's mercy will be poured forth only for a certain time, my son, until the majority of my prophecies are fulfilled and this fulfillment is complete. Then several times in the course of the doleful tempests the Lord shall say: Therefore I shall crush and destroy and show no mercy; and many other circumstances shall result from floods and continual rain of which I have written more fully in my other prophecies, composed at some length, not in a chronological sequence, in prose, limiting the places and times and exact dates so that future generations will see, while experiencing these inevitable events, how I have listed others in clearer language, so that despite their obscurities these things shall be understood: When the time comes for the removal of ignorance, the matter will be clearer still.

So in conclusion, my son, take this gift from your father M. Nostradamus, who hopes you will understand each prophecy in every quatrain herein. May Immortal God grant you a long life of good and prosperous happiness.

Salon, 1 March 1555

Century I

1
Sitting alone at night in secret study;
it is placed on the brass tripod.
A slight flame comes out of the emptiness and
makes successful that which should not be believed in vain.

2
The wand in the hand is placed in the middle of the tripod's legs.
With water he sprinkles both the hem of his garment and his foot.
A voice, fear: he trembles in his robes.
Divine splendor; the God sits nearby.

3
When the litters are overturned by the whirlwind
and faces are covered by cloaks,
the new republic will be troubled by its people.
At this time the reds and the whites will rule wrongly.

4
In the world there will be made a king
who will have little peace and a short life.
At this time the ship of the Papacy will be lost,
governed to its greatest detriment.

5
They will be driven away for a long drawn out fight.
The countryside will be most grievously troubled.
Town and country will have greater struggle.
Carcassonne and Narbonne will have their hearts tried.

6
The eye of Ravenna will be forsaken,
when his wings will fail at his feet.
The two of Bresse will have made a constitution
for Turin and Vercelli, which the French will trample underfoot

7
Arrived too late, the act has been done.
The wind was against them, letters intercepted on their way.

The conspirators were fourteen of a party.
By Rousseau shall these enterprises be undertaken.

8

How often will you be captured, O city of the sun ?
Changing laws that are barbaric and vain.
Bad times approach you. No longer will you be enslaved.
Great Hadrie will revive your veins.

9

From the Orient will come the African heart
to trouble Hadrie and the heirs of Romulus.
Accompanied by the Libyan fleet
the temples of Malta and nearby islands shall be deserted.

10

A coffin is put into the vault of iron,
where seven children of the king are held.
The ancestors and forebears will come forth from the depths of hell,
lamenting to see thus dead the fruit of their line.

11

The motion of senses, heart, feet and hands
will be in agreement between Naples, Lyon and Sicily.
Swords fire, floods, then the noble Romans drowned,
killed or dead because of a weak brain.

12

There will soon be talk of a treacherous man, who rules a short time,
quickly raised from low to high estate.
He will suddenly turn disloyal and volatile.
This man will govern Verona.

13

Through anger and internal hatreds, the exiles
will hatch a great plot against the king.
Secretly they will place enemies as a threat,
and his own old (adherents) will find sedition against them.

14

From the enslaved populace, songs, chants and demands,
while Princes and Lords are held captive in prisons.
These will in the future by headless idiots
be received as divine prayers

15

.Mars threatens us with the force of war
and will cause blood to be spilt seventy times.
The clergy will be both exalted and reviled moreover,
by those who wish to learn nothing of them.

16
A scythe joined with a pond in Sagittarius
at its highest ascendant.
Plague, famine, death from military hands;
the century approaches its renewal.

17
For forty years the rainbow will not be seen.
For forty years it will be seen every day.
The dry earth will grow more parched,
and there will be great floods when it is seen.

18
Because of French discord and negligence
an opening shall be given to the Mohammedans.
The land and sea of Siena will be soaked in blood,
and the port of Marseilles covered with ships and sails.

19
When the snakes surround the altar,
and the Trojan blood is troubled by the Spanish.
Because of them, a great number will be lessened.
The leader flees, hidden in the swampy marshes.

20
The cities of Tours, Orleans, Blois, Angers, Reims and Nantes
are troubled by sudden change.
Tents will be pitched by (people) of foreign tongues;
rivers, darts at Rennes, shaking of land and sea.

21
The rock holds in its depths white clay
which will come out milk–white from a cleft
Needlessly troubled people will not dare touch it,
unaware that the foundation of the earth is of clay.

22
A thing existing without any senses
will cause its own end to happen through artifice.
At Autun, Chalan, Langres and the two Sens
there will be great damage from hail and ice.

23
In the third month, at sunrise,
the Boar and the Leopard meet on the battlefield.
The fatigued Leopard looks up to heaven
and sees an eagle playing around the sun.

24
At the New City he is thoughtful to condemn;
the bird of prey offers himself to the Gods.

After victory he pardons his captives.
At Cremona and Mantua great hardships will be suffered.

25
The lost thing is discovered, hidden for many centuries.
Pasteur will be celebrated almost as a God–like figure.
This is when the moon completes her great cycle,
but by other rumors he shall be dishonored.

26
The great man will be struck down in the day by a thunderbolt.
An evil deed, foretold by the bearer of a petition.
According to the prediction another falls at night time.
Conflict at Reims, London, and pestilence in Tuscany.

27
Beneath the oak tree of Gienne, struck by lightning,
the treasure is hidden not far from there.
That which for many centuries had been gathered,
when found, a man will die, his eye pierced by a spring.

28
Tobruk will fear the barbarian fleet for a time,
then much later the Western fleet.
Cattle, people, possessions, all will be quite lost.
What a deadly combat in Taurus and Libra.

29
When the fish that travels over both land and sea
is cast up on to the shore by a great wave,
its shape foreign, smooth and frightful.
From the sea the enemies soon reach the walls.

30
Because of the storm at sea the foreign ship
will approach an unknown port.
Notwithstanding the signs of the palm branches,
afterwards there is death and pillage. Good advice comes too late.

31
The wars in France will last for so many years
beyond the reign of the Castulon kings.
An uncertain victory will crown three great ones,
the Eagle, the Cock, the Moon, the Lion, the Sun in its house.

32
The great Empire will soon be exchanged
for a small place, which soon will begin to grow.
A small place of tiny area
in the middle of which he will come to lay down his scepter.

33
Near a great bridge near a spacious plain
the great lion with the Imperial forces
will cause a falling outside the austere city.
Through fear the gates will be unlocked for him.

34
The bird of prey flying to the left,
before battle is joined with the French, he makes preparations.
Some will regard him as good, others bad or uncertain.
The weaker party will regard him as a good omen.

35
The young lion will overcome the older one,
in a field of combat in single fight:
He will pierce his eyes in their golden cage;
two wounds in one, then he dies a cruel death.

36
Too late the king will repent
that he did not put his adversary to death.
But he will soon come to agree to far greater things
which will cause all his line to die.

37
Shortly before sun set, battle is engaged.
A great nation is uncertain.
Overcome, the sea port makes no answer,
the bridge and the grave both in foreign places.

38
The Sun and the Eagle will appear to the victor.
An empty answer assured to the defeated.
Neither bugle nor shouts will stop the soldiers.
Liberty and peace, if achieved in time through death.

39
At night the last one will be strangled in his bed
because he became too involved with the blond heir elect.
The Empire is enslaved and three men substituted.
He is put to death with neither letter nor packet read.

40
The false trumpet concealing madness
will cause Byzantium to change its laws.
From Egypt there will go forth a man who wants
the edict withdrawn, changing money and standards.

41
The city is besieged and assaulted by night;
few have escaped; a battle not far from the sea.

A woman faints with joy at the return of her son,
poison in the folds of the hidden letters.

42
The tenth day of the April Calends, calculated in Gothic fashion
is revived again by wicked people.
The fire is put out and the diabolic gathering
seek the bones of the demon of Psellus.

43
Before the Empire changes
a very wonderful event will take place.
The field moved, the pillar of porphyry
put in place, changed on the gnarled rock.

44
In a short time sacrifices will be resumed,
those opposed will be put (to death) like martyrs.
The will no longer be monks, abbots or novices.
Honey shall be far more expensive than wax.

45
A founder of sects, much trouble for the accuser:
A beast in the theater prepares the scene and plot.
The author ennobled by acts of older times;
the world is confused by schismatic sects.

46
Very near Auch, Lectoure and Mirande
a great fire will fall from the sky for three nights.
The cause will appear both stupefying and marvelous;
shortly afterwards there will be an earthquake.

47
The speeches of Lake Leman will become angered,
the days will drag out into weeks,
then months, then years, then all will fail.
The authorities will condemn their useless powers.

48
When twenty years of the Moon's reign have passed
another will take up his reign for seven thousand years.
When the exhausted Sun takes up his cycle
then my prophecy and threats will be accomplished.

49
Long before these happenings
the people of the East, influenced by the Moon,
in the year 1700 will cause many to be carried away,
and will almost subdue the Northern area.

50
From the three water signs will be born a man
who will celebrate Thursday as his holiday.
His renown, praise, rule and power will grow
on land and sea, bringing trouble to the East.

51
The head of Aries, Jupiter and Saturn.
Eternal God, what changes !
Then the bad times will return again after a long century;
what turmoil in France and Italy.

52
Two evil influences in conjunction in Scorpio.
The great lord is murdered in his room.
A newly appointed king persecutes the Church,
the lower (parts of) Europe and in the North.

53
Alas, how we will see a great nation sorely troubled
and the holy law in utter ruin.
Christianity (governed) throughout by other laws,
when a new source of gold and silver is discovered.

54
Two revolutions will be caused by the evil scythe bearer
making a change of reign and centuries.
The mobile sign thus moves into its house:
Equal in favor to both sides.

55
In the land with a climate opposite to Babylon
there will be great shedding of blood.
Heaven will seem unjust both on land and sea and in the air.
Sects, famine, kingdoms, plagues, confusion.

56
Sooner and later you will see great changes made,
dreadful horrors and vengeances.
For as the moon is thus led by its angel
the heavens draw near to the Balance.

57
The trumpet shakes with great discord.
An agreement broken: lifting the face to heaven:
the bloody mouth will swim with blood;
the face anointed with milk and honey lies on the ground.

58
Through a slit in the belly a creature will be born with two heads
and four arms: it will survive for some few years.

The day that Alquiloie celebrates his festivals
Fossana, Turin and the ruler of Ferrara will follow.

59
The exiles deported to the islands
at the advent of an even more cruel king
will be murdered. Two will be burnt
who were not sparing in their speech.

60
An Emperor will be born near Italy,
who will cost the Empire very dearly.
They will say, when they see his allies,
that he is less a prince than a butcher.

61
The wretched, unfortunate republic
will again be ruined by a new authority.
The great amount of ill will accumulated in exile
will make the Swiss break their important agreement.

62
Alas! what a great loss there will be to learning
before the cycle of the Moon is completed.
Fire, great floods, by more ignorant rulers;
how long the centuries until it is seen to be restored.

63
Pestilences extinguished, the world becomes smaller,
for a long time the lands will be inhabited peacefully.
People will travel safely through the sky (over) land and seas:
then wars will start up again.

64
At night they will think they have seen the sun,
when the see the half pig man:
Noise, screams, battles seen fought in the skies.
The brute beasts will be heard to speak.

65
A child without hands, never so great a thunderbolt seen,
the royal child wounded at a game of tennis.
At the well lightning strikes, joining together
three trussed up in the middle under the oaks.

66
He who then carries the news,
after a short while will (stop) to breathe:
Viviers, Tournon, Montferrand and Praddelles;
hail and storms will make them grieve.

67
The great famine which I sense approaching
will often turn (in various areas) then become worldwide.
It will be so vast and long lasting that (they) will grab
roots from the trees and children from the breast.

68
O to what a dreadful and wretched torment
are three innocent people going to be delivered.
Poison suggested, badly guarded, betrayal.
Delivered up to horror by drunken executioners.

69
The great mountain, seven stadia round,
after peace, war, famine, flooding.
It will spread far, drowning great countries,
even antiquities and their mighty foundations.

70
Rain, famine and war will not cease in Persia;
too great a faith will betray the monarch.
Those (actions) started in France will end there,
a secret sign for on to be sparing.

71
The marine tower will be captured and retaken three times
by Spaniards, Barbarians and Ligurians.
Marseilles and Aix, Ales by men of Pisa,
devastation, fire, sword, pillage at Avignon by the Turinese.

72
The inhabitants of Marseilles completely changed,
fleeing and pursued as far as Lyons.
Narbonne, Toulouse angered by Bordeaux;
the killed and captive are almost one million.

73
France shall be accused of neglect by her five partners.
Tunis, Algiers stirred up by the Persians.
Leon, Seville and Barcelona having failed,
they will not have the fleet because of the Venetians.

74
After a rest they will travel to Epirus,
great help coming from around Antioch.
The curly haired king will strive greatly for the Empire,
the brazen beard will be roasted on a spit.

75
The tyrant of Siena will occupy Savona,
having won the fort he will restrain the marine fleet.

Two armies under the standard of Ancona:
the leader will examine them in fear.

76
The man will be called by a barbaric name
that three sisters will receive from destiny.
He will speak then to a great people in words and deeds,
more than any other man will have fame and renown.

77
A promontory stands between two seas:
A man who will die later by the bit of a horse;
Neptune unfurls a black sail for his man;
the fleet near Gibraltar and Rocheval.

78
To an old leader will be born an idiot heir,
weak both in knowledge and in war.
The leader of France is feared by his sister,
battlefields divided, conceded to the soldiers.

79
Bazas, Lectoure, Condom, Auch and Agen
are troubled by laws, disputes and monopolies.
Carcassone, Bordeaux, Toulouse and Bayonne will be ruined
when they wish to renew the massacre.

80
From the sixth bright celestial light
it will come to thunder very strongly in Burgundy.
Then a monster will be born of a very hideous beast:
In March, April, May and June great wounding and worrying.

81
Nine will be set apart from the human flock,
separated from judgment and advise.
Their fate is to be divided as they depart.
K. Th. L. dead, banished and scattered.

82
When the great wooden columns tremble
in the south wind, covered with blood.
Such a great assembly then pours forth
that Vienna and the land of Austria will tremble.

83
The alien nation will divide the spoils.
Saturn in dreadful aspect in Mars.
Dreadful and foreign to the Tuscans and Latins,
Greeks who will wish to strike.

84
The moon is obscured in deep gloom,
his brother becomes bright red in color.
The great one hidden for a long time in the shadows
will hold the blade in the bloody wound.

85
The king is troubled by the queen's reply.
Ambassadors will fear for their lives.
The greater of his brothers will doubly disguise his action,
two of them will die through anger, hatred and envy.

86
When the great queen sees herself conquered,
she will show an excess of masculine courage.
Naked, on horseback, she will pass over the river
pursued by the sword: she will have outraged her faith

87
Earthshaking fire from the center of the earth
will cause tremors around the New City.
Two great rocks will war for a long time,
then Arethusa will redden a new river.

88
The divine wrath overtakes the great Prince,
a short while before he will marry.
Both supporters and credit will suddenly diminish.
Counsel, he will die because of the shaven heads.

89
Those of Lerida will be in the Moselle,
kill all those from the Loire and Seine.
The seaside track will come near the high valley,
when the Spanish open every route.

90
Bordeaux and Poitiers at the sound of the bell
will go with a great fleet as fast as Langon.
A great rage will surge up against the French,
when a hideous monster is born near Orgon.

91
The Gods will make it appear to mankind
that they are the authors of a great war.
Before the sky was seen to bee free of weapons and rockets:
the greatest damage will be inflicted on the left.

92
Under one man peace will be proclaimed everywhere,
but not long after will be looting and rebellion.

Because of a refusal, town, land and see will be broached.
About a third of a million dead or captured.

93
The Italian lands near the mountains will tremble.
The Cock and the Lion not strongly united.
In place of fear they will help each other.
Freedom alone moderates the French.

94
The tyrant Selim will be put to death at the harbor
but Liberty will not be regained, however.
A new war arises from vengeance and remorse.
A lady is honored through force of terror.

95
In front of a monastery will be found a twin infant
from the illustrious and ancient line of a monk.
His fame, renown and power through sects and speech
is such that they will say the living twin is deservedly chosen.

96
A man will be charged with the destruction
of temples and sects, altered by fantasy.
He will harm the rocks rather than the living,
ears filled with ornate speeches.

97
That which neither weapon nor flame could accomplish
will be achieved by a sweet speaking tongue in council.
Sleeping, in a dream, the king will see
the enemy not in war or of military blood.

98
The leader who will conduct great numbers of people
far from their skies, to foreign customs and language.
Five thousand will die in Crete and Thessaly,
the leader fleeing in a sea going supply ship.

99
The great king will join
with two kings, united in friendship.
How the great household will sigh:
around Narbon what pity for the children.

100
For a long time a gray bird will be seen in the sky
near Dile and the lands of Tuscany.
He holds a flowering branch in his beak,
but he dies too soon and the war ends.

Century II

1
Towards Aquitaine by the British Isles
By these themselves great incursions.
Rains, frosts will make the soil uneven,
Port Selyn will make mighty invasions

2
The blue head will inflict upon the white head
As much evil as France has done them good:
Dead at the sail-yard the great one hung on the branch.
When seized by his own the King will say how much.

3
Because of the solar heat on the sea
From Negrepont the fishes half cooked:
The inhabitants will come to cut them,
When food will fail in Rhodes and Genoa.

4
From Monaco to near Sicily
The entire coast will remain desolated:
There will remain there no suburb, city or town
Not pillaged and robbed by the Barbarians.

5
That which is enclosed in iron and letter in a fish,
Out will go one who will then make war,
He will have his fleet well rowed by sea,
Appearing near Latin land.

6
Near the gates and within two cities
There will be two scourges the like of which was never seen,
Famine within plague, people put out by steel,
Crying to the great immortal God for relief.

7
Amongst several transported to the isles,
One to be born with two teeth in his mouth
They will die of famine the trees stripped,
For them a new King issues a new edict.

8
Temples consecrated in the original Roman manner,
They will reject the excess foundations,
Taking their first and humane laws,
Chasing, though not entirely, the cult of saints.

9
Nine years the lean one will hold the realm in peace,
Then he will fall into a very bloody thirst:
Because of him a great people will die without faith and law
Killed by one far more good-natured.

10
Before long all will be set in order,
We will expect a very sinister century,
The state of the masked and solitary ones much changed,
Few will be found who want to be in their place.

11
The nearest son of the elder will attain
Very great height as far as the realm of the privileged:
Everyone will fear his fierce glory,
But his children will be thrown out of the realm.

12
Eyes closed, opened by antique fantasy,
The garb of the monks they will be put to naught:
The great monarch will chastise their frenzy,
Ravishing the treasure in front of the temples.

13
The body without soul no longer to be sacrificed:
Day of death put for birthday:
The divine spirit will make the soul happy,
Seeing the word in its eternity.

14
At Tours, Gien, guarded, eyes will be searching,
Discovering from afar her serene Highness:
She and her suite will enter the port,
Combat, thrust, sovereign power.

15
Shortly before the monarch is assassinated,
Castor and Pollux in the ship, bearded star:
The public treasure emptied by land and sea,
Pisa, Asti, Ferrara, Turin land under interdict.

16
Naples, Palermo, Sicily, Syracuse,
New tyrants, celestial lightning fires:
Force from London, Ghent, Brussels and Susa,
Great slaughter, triumph leads to festivities.

17
The field of the temple of the vestal virgin,
Not far from Elne and the Pyrenees mountains:

The great tube is hidden in the trunk.
To the north rivers overflown and vines battered.

18
New, impetuous and sudden rain
Will suddenly halt two armies.
Celestial stone, fires make the sea stony,
The death of seven by land and sea sudden.

19
Newcomers, place built without defense,
Place occupied then uninhabitable:
Meadows, houses, fields, towns to take at pleasure,
Famine, plague, war, extensive land arable.

20
Brothers and sisters captive in diverse places
Will find themselves passing near the monarch:
Contemplating them his branches attentive,
Displeasing to see the marks on chin, forehead and nose.

21
The ambassador sent by biremes,
Halfway repelled by unknown ones:
Reinforced with salt four triremes will come,
In Euboea bound with ropes and chains.

22
The imprudent army of Europe will depart,
Collecting itself near the submerged isle:
The weak fleet will bend the phalanx,
At the navel of the world a greater voice substituted.

23
Palace birds, chased out by a bird,
Very soon after the prince has arrived:
Although the enemy is repelled beyond the river,
Outside seized the trick upheld by the bird.

24
Beasts ferocious from hunger will swim across rivers:
The greater part of the region will be against the Hister,
The great one will cause it to be dragged in an iron cage,
When the German child will observe nothing.

25
The foreign guard will betray the fortress,
Hope and shadow of a higher marriage:
Guard deceived, fort seized in the press,
Loire, Saone, Rhone, Garonne, mortal outrage.

26
Because of the favor that the city will show
To the great one who will soon lose the field of battle,
Fleeing the Po position, the Ticino will overflow
With blood, fires, deaths, drowned by the long-edged blow.

27
The divine word will be struck from the sky,
One who cannot proceed any further:
The secret closed up with the revelation,
Such that they will march over and ahead.

28
The penultimate of the surname of the Prophet
Will take Diana [Thursday] for his day and rest:
He will wander far because of a frantic head,
And delivering a great people from subjection.

29
The Easterner will leave his seat,
To pass the Apennine mountains to see Gaul:
He will transpire the sky, the waters and the snow,
And everyone will be struck with his rod.

30
One who the infernal gods of Hannibal
Will cause to be reborn, terror of mankind
Never more horror nor worse of days
In the past than will come to the Romans through Babel.

31
In Campania the Capuan [river] will do so much
That one will see only fields covered by waters:
Before and after the long rain
One will see nothing green except the trees.

32
Milk, frog's blood prepared in Dalmatia.
Conflict given, plague near Treglia:
A great cry will sound through all Slavonia,
Then a monster will be born near and within Ravenna.

33
Through the torrent which descends from Verona
Its entry will then be guided to the Po,
A great wreck, and no less in the Garonne,
When those of Genoa march against their country.

34
The senseless ire of the furious combat
Will cause steel to be flashed at the table by brothers:

To part them death, wound, and curiously,
The proud duel will come to harm France.

35
The fire by night will take hold in two lodgings,
Several within suffocated and roasted.
It will happen near two rivers as one:
Sun, Sagittarius and Capricorn all will be reduced.

36
The letters of the great Prophet will be seized,
They will come to fall into the hands of the tyrant:
His enterprise will be to deceive his King,
But his extortions will very soon trouble him.

37
Of that great number that one will send
To relieve those besieged in the fort,
Plague and famine will devour them all,
Except seventy who will be destroyed.

38
A great number will be condemned
When the monarchs will be reconciled:
But for one of them such a bad impediment will arise
That they will be joined together but loosely.

39.
One year before the Italian conflict,
Germans, Gauls, Spaniards for the fort:
The republican schoolhouse will fall,
There, except for a few, they will be choked dead.

40
Shortly afterwards, without a very long interval,
By sea and land a great uproar will be raised:
Naval battle will be very much greater,
Fires, animals, those who will cause greater insult.

41
The great star will burn for seven days,
The cloud will cause two suns to appear:
The big mastiff will howl all night
When the great pontiff will change country.

42
Cock, dogs and cats will be satiated with blood
And from the wound of the tyrant found dead,
At the bed of another legs and arms broken,
He who was not afraid to die a cruel death.

43
During the appearance of the bearded star.
The three great princes will be made enemies:
Struck from the sky, peace earth quaking,
Po, Tiber overflowing, serpent placed upon the shore.

44
The Eagle driven back around the tents
Will be chased from there by other birds:
When the noise of cymbals, trumpets and bells
Will restore the senses of the senseless lady.

45.
Too much the heavens weep for the Androgyne begotten,
Near the heavens human blood shed:
Because of death too late a great people re-created,
Late and soon the awaited relief comes.

46
After great trouble for humanity, a greater one is prepared
The Great Mover renews the ages:
Rain, blood, milk, famine, steel and plague,
Is the heavens fire seen, a long spark running.

47
The great old enemy mourning dies of poison,
The sovereigns subjugated in infinite numbers:
Stones raining, hidden under the fleece,
Through death articles are cited in vain.

48
The great force which will pass the mountains.
Saturn in Sagittarius Mars turning from the fish:
Poison hidden under the heads of salmon,
Their war-chief hung with cord.

49
The advisers of the first monopoly,
The conquerors seduced for Malta:
Rhodes, Byzantium for them exposing their pole:
Land will fail the pursuers in flight.

50
When those of Hainault, of Ghent and of Brussels
Will see the siege laid before Langres:
Behind their flanks there will be cruel wars,
The ancient wound will do worse than enemies.

52
The blood of the just will commit a fault at London,
Burnt through lightning of twenty threes the six:

The ancient lady will fall from her high place,
Several of the same sect will be killed.

52
For several nights the earth will tremble:
In the spring two efforts in succession:
Corinth, Ephesus will swim in the two seas:
War stirred up by two valiant in combat.

53
The great plague of the maritime city
Will not cease until there be avenged the death
Of the just blood, condemned for a price without crime,
Of the great lady outraged by pretense.

54.
Because of people strange, and distant from the Romans
Their great city much troubled after water:
Daughter handless, domain too different,
Chief taken, lock not having been picked.

55
In the conflict the great one who was worth little
At his end will perform a marvelous deed:
While Adria will see what he was lacking,
During the banquet the proud one stabbed.

56
One whom neither plague nor steel knew how to finish,
Death on the summit of the hills struck from the sky:
The abbot will die when he will see ruined
Those of the wreck wishing to seize the rock.

57
Before the conflict the great wall will fall,
The great one to death, death too sudden and lamented,
Born imperfect: the greater part will swim:
Near the river the land stained with blood.

58
With neither foot nor hand because of sharp and strong tooth
Through the crowd to the fort of the pork and the elder born:
Near the portal treacherous proceeds,
Moon shining, little great one led off.

59
Gallic fleet through support of the great guard
Of the great Neptune, and his trident soldiers,
Provence reddened to sustain a great band:
More at Narbonne, because of javelins and darts.

60
The Punic faith broken in the East,
Ganges, Jordan, and Rhone, Loire, and Tagus will change:
When the hunger of the mule will be satiated,
Fleet sprinkles, blood and bodies will swim.

61
Bravo, ye of Tamins, Gironde and La Rochelle:
O Trojan blood! Mars at the port of the arrow
Behind the river the ladder put to the fort,
Points to fire great murder on the breach.

62
Mabus then will soon die, there will come
Of people and beasts a horrible rout:
Then suddenly one will see vengeance,
Hundred, hand, thirst, hunger when the comet will run.

64
The Gauls Ausonia will subjugate very little,
Po, Marne and Seine Parma will make drunk:
He who will prepare the great wall against them,
He will lose his life from the least at the wall.

64
The people of Geneva drying up with hunger, with thirst,
Hope at hand will come to fail:
On the point of trembling will be the law of him of the Cevennes,
Fleet at the great port cannot be received.

65
The sloping park great calamity
To be done through Hesperia and Insubria:
The fire in the ship, plague and captivity,
Mercury in Sagittarius Saturn will fade.

66
Through great dangers the captive escaped:
In a short time great his fortune changed.
In the palace the people are trapped,
Through good omen the city besieged.

67
The blond one will come to compromise the fork-nosed one
Through the duel and will chase him out:
The exiles within he will have restored,
Committing the strongest to the marine places.

68
The efforts of Aquilon will be great:
The gate on the Ocean will be opened,

The kingdom on the Isle will be restored:
London will tremble discovered by sail.

69
The Gallic King through his Celtic right arm
Seeing the discord of the great Monarchy:
He will cause his scepter to flourish over the three parts,
Against the cope of the great Hierarchy.

70
The dart from the sky will make its extension,
Deaths speaking: great execution.
The stone in the tree, the proud nation restored,
Noise, human monster, purge expiation.

71
The exiles will come into Sicily
To deliver form hunger the strange nation:
At daybreak the Celts will fail them:
Life remains by reason: the King joins.

72
Celtic army vexed in Italy
On all sides conflict and great loss:
Romans fled, O Gaul repelled!
Near the Ticino, Rubicon uncertain battle.

73
The shore of Lake Garda to Lake Fucino,
Taken from the Lake of Geneva to the port of L'Orguion:
Born with three arms the predicted warlike image,
Through three crowns to the great Endymion.

74
From Sens, from Autun they will come as far as the Rhone
To pass beyond towards the Pyrenees mountains:
The nation to leave the March of Ancona:
By land and sea it will be followed by great suites.

75
On the pipe of the air-vent floor:
So high will the bushel of wheat rise,
That man will be eating his fellow man.

76.
Lightning in Burgundy will perform a portentous deed,
One which could never have been done by skill,
Sexton made lame by their senate
Will make the affair known to the enemies.

77
Hurled back through bows, fires, pitch and by fires:
Cries, howls heard at midnight:
Within they are place on the broken ramparts,
The traitors fled by the underground passages.

78.
The great Neptune of the deep of the sea
With Punic race and Gallic blood mixed.
The Isles bled, because of the tardy rowing:
More harm will it do him than the ill-concealed secret.

79
The beard frizzled and black through skill
Will subjugate the cruel and proud people:
The great Chyren will remove from far away
All those captured by the banner of Selin

80
After the conflict by the eloquence of the wounded one
For a short time a soft rest is contrived:
The great ones are not to be allowed deliverance at all:
They are restored by the enemies at the proper time.

81
Through fire from the sky the city almost burned:
The Urn threatens Deucalion again:
Sardinia vexed by the Punic foist,
After Libra will leave her Phaethon.

82
Through hunger the prey will make the wolf prisoner,
The aggressor then in extreme distress.
The heir having the last one before him,
The great one does not escape in the middle of the crowd.

83
The large trade of a great Lyons changed,
The greater part turns to pristine ruin
Prey to the soldiers swept away by pillage:
Through the Jura mountain and Suevia drizzle.

84
Between Campania, Siena, Florence, Tuscany,
Six months nine days without a drop of rain:
The strange tongue in the Dalmatian land,
It will overrun, devastating the entire land.

85
The old full beard under the severe statute
Made at Lyon over the Celtic Eagle:

The little great one perseveres too far:
Noise of arms in the sky: Ligurian sea red.

86
Wreck for the fleet near the Adriatic Sea:
The land trembles stirred up upon the air placed on land:
Egypt trembles Mahometan increase,
The Herald surrendering himself is appointed to cry out.

87
After there will come from the outermost countries
A German Prince, upon the golden throne:
The servitude and waters met,
The lady serves, her time no longer adored.

88
The circuit of the great ruinous deed,
The seventh name of the fifth will be:
Of a third greater the stranger warlike:
Sheep, Paris, Aix will not guarantee.

89
One day the two great masters will be friends,
Their great power will be seen increased:
The new land will be at its high peak,
To the bloody one the number recounted.

90
Though life and death the realm of Hungary changed:
The law will be more harsh than service:
Their great city cries out with howls and laments,
Castor and Pollux enemies in the arena.

91.
At sunrise one will see a great fire,
Noise and light extending towards Aquilon:
Within the circle death and one will hear cries,
Through steel, fire, famine, death awaiting them.

92
Fire color of gold from the sky seen on earth:
Heir struck from on high, marvelous deed done:
Great human murder: the nephew of the great one taken,
Deaths spectacular the proud one escaped.

93
Very near the Tiber presses Death:
Shortly before great inundation:
The chief of the ship taken, thrown into the bilge:
Castle, palace in conflagration.

94
Great Po, great evil will be received through Gauls,
Vain terror to the maritime Lion:
People will pass by the sea in infinite numbers,
Without a quarter of a million escaping.

95
The populous places will be uninhabitable:
Great discord to obtain fields:
Realms delivered to prudent incapable ones:
Then for the great brothers dissension and death.

96
Burning torch will be seen in the sky at night
Near the end and beginning of the Rhone:
Famine, steel: the relief provided late,
Persia turns to invade Macedonia.

97
Roman Pontiff beware of approaching
The city that two rivers flow through,
Near there your blood will come to spurt,
You and yours when the rose will flourish.

98
The one whose face is splattered with the blood
Of the victim nearly sacrificed:
Jupiter in Leon, omen through presage:
To be put to death then for the bride.

99
Roman land as the omen interpreted
Will be vexed too much by the Gallic people:
But the Celtic nation will fear the hour,
The fleet has been pushed too far by the north wind.

100
Within the isles a very horrible uproar,
One will hear only a party of war,
So great will be the insult of the plunderers
That they will come to be joined in the great league.

Century III

1
After combat and naval battle,
The great Neptune in his highest belfry:
Red adversary will become pale with fear,
Putting the great Ocean in dread.

2
The divine word will give to the sustenance,
Including heaven, earth, gold hidden in the mystic milk:
Body, soul, spirit having all power,
As much under its feet as the Heavenly see.

3
Mars and Mercury, and the silver joined together,
Towards the south extreme drought:
In the depths of Asia one will say the earth trembles,
Corinth, Ephesus then in perplexity.

4
When they will be close the lunar ones will fail,
From one another not greatly distant,
Cold, dryness, danger towards the frontiers,
Even where the oracle has had its beginning.

5
Near, far the failure of the two great luminaries
Which will occur between April and March.
Oh, what a loss! but two great good-natured ones
By land and sea will relieve all parts.

6
Within the closed temple the lightning will enter,
The citizens within their fort injured:
Horses, cattle, men, the wave will touch the wall,
Through famine, drought, under the weakest armed.

7
The fugitives, fire from the sky on the pikes:
Conflict near the ravens frolicking,
From land they cry for aid and heavenly relief,
When the combatants will be near the walls.

8
The Cimbri joined with their neighbors
Will come to ravage almost Spain:
Peoples gathered in Guienne and Limousin
Will be in league, and will bear them company.

9
Bordeaux, Rouen and La Rochelle joined
Will hold around the great Ocean sea,
English, Bretons and the Flemings allied
Will chase them as far as Roanne.

10
Greater calamity of blood and famine,
Seven times it approaches the marine shore:

Monaco from hunger, place captured, captivity,
The great one led crunching in a metaled cage.

11
The arms to fight in the sky a long time,
The tree in the middle of the city fallen:
Sacred bough clipped, steel, in the face of the firebrand,
Then the monarch of Adria fallen.

12
Because of the swelling of the Ebro, Po, Tagus, Tiber and Rhine
And because of the pond of Geneva and Arezzo,
The two great chiefs and cities of the Garonne,
Taken, dead, drowned: human booty divided.

13.
Through lightning in the arch gold and silver melted,
Of two captives one will eat the other:
The greatest one of the city stretched out,
When submerged the fleet will swim.

14
Through the branch of the valiant personage
Of lowest France: because of the unhappy father
Honors, riches, travail in his old age,
For having believed the advice of a simple man.

15
The realm, will change in heart, vigor and glory,
In all points having its adversary opposed:
Then through death France an infancy will subjugate,
A great Regent will then be more contrary.

16
An English prince Marc in his heavenly heart
Will want to pursue his prosperous fortune,
Of the two duels one will pierce his gall:
Hated by him well loved by his mother.

17
Mount Aventine will be seen to burn at night:
The sky very suddenly dark in Flanders:
When the monarch will chase his nephew,
Then Church people will commit scandals.

18
After the rather long rain milk,
In several places in Reims the sky touched:
Alas, what a bloody murder is prepared near them,
Fathers and sons Kings will not dare approach.

19
In Lucca it will come to rain blood and milk,
Shortly before a change of praetor:
Great plague and war, famine and drought will be made visible
Far away where their prince and rector will die.

20
Through the regions of the great river Guadalquivir
Deep in Iberia to the Kingdom of Grenada
Crosses beaten back by the Mahometan peoples
One of Cordova will betray his country

21
In the Conca by the Adriatic Sea
There will appear a horrible fish,
With face human and its end aquatic,
Which will be taken without the hook.

22
Six days the attack made before the city:
Battle will be given strong and harsh:
Three will surrender it, and to them pardon:
The rest to fire and to bloody slicing and cutting.

23
If, France, you pass beyond the Ligurian Sea,
You will see yourself shut up in islands and seas:
Mahomet contrary, more so the Adriatic Sea:
You will gnaw the bones of horses and asses.

24
Great confusion in the enterprise,
Loss of people, countless treasure:
You ought not to extend further there.
France, let what I say be remembered.

25
He who will attain to the kingdom of Navarre
When Sicily and Naples will be joined:
He will hold Bigorre and Landes through Foix and Oloron
From one who will be too closely allied with Spain.

26
They will prepare idols of Kings and Princes,
Soothsayers and empty prophets elevated:
Horn, victim of gold, and azure, dazzling,
The soothsayers will be interpreted.

27
Libyan Prince powerful in the West
Will come to inflame very much French with Arabian.

Learned in letters condescending he will
Translate the Arabian language into French.

28
Of land weak and parentage poor,
Through piece and peace he will attain to the empire.
For a long time a young female to reign,
Never has one so bad come upon the kingdom.

29
The two nephews brought up in diverse places:
Naval battle, land, fathers fallen:
They will come to be elevated very high in making war
To avenge the injury, enemies succumbed.

30
He who during the struggle with steel in the deed of war
Will have carried off the prize from on greater than he:
By night six will carry the grudge to his bed,
Without armor he will surprised suddenly.

31
On the field of Media, of Arabia and of Armenia
Two great armies will assemble thrice:
The host near the bank of the Araxes,
They will fall in the land of the great Suleiman.

32
The great tomb of the people of Aquitaine
Will approach near to Tuscany,
When Mars will be in the corner of Germany
And in the land of the Mantuan people.

33
In the city where the wolf will enter,
Very near there will the enemies be:
Foreign army will spoil a great country.
The friends will pass at the wall and Alps.

34
When the eclipse of the Sun will then be,
The monster will be seen in full day:
Quite otherwise will one interpret it,
High price unguarded: none will have foreseen it.

35
From the very depths of the West of Europe,
A young child will be born of poor people,
He who by his tongue will seduce a great troop:
His fame will increase towards the realm of the East.

36
Buried apoplectic not dead,
He will be found to have his hands eaten:
When the city will condemn the heretic,
He who it seemed to them had changed their laws.

37
The speech delivered before the attack,
Milan taken by the Eagle through deceptive ambushes:
Ancient wall driven in by cannons,
Through fire and blood few given quarter.

38
The Gallic people and a foreign nation
Beyond the mountains, dead, captured and killed:
In the contrary month and near vintage time,
Through the Lords drawn up in accord.

39
The seven in three months in agreement
To subjugate the Apennine Alps:
But the tempest and cowardly Ligurian,
Destroys them in sudden ruins.

40
The great theater will come to be set up again:
The dice cast and the snares already laid.
Too much the first one will come to tire in the death knell,
Prostrated by arches already a long time split.

41
Hunchback will be elected by the council,
A more hideous monster not seen on earth,
The willing blow will put out his eye:
The traitor to the King received as faithful.

42
The child will be born with two teeth in his mouth,
Stones will fall during the rain in Tuscany:
A few years after there will be neither wheat nor barley,
To satiate those who will faint from hunger.

43
People from around the Tarn, Lot and Garonne
Beware of passing the Apennine mountains:
Your tomb near Rome and Ancona,
The black frizzled beard will have a trophy set up.

44
When the animal domesticated by man
After great pains and leaps will come to speak:

The lightning to the virgin will be very harmful,
Taken from earth and suspended in the air.

45
The five strangers entered in the temple,
Their blood will come to pollute the land:
To the Toulousans it will be a very hard example
Of one who will come to exterminate their laws.

46
The sky (of Plancus' city) forebodes to us
Through clear signs and fixed stars,
That the time of its sudden change is approaching,
Neither for its good, nor for its evils.

47
The old monarch chased out of his realm
Will go to the East asking for its help:
For fear of the crosses he will fold his banner:
To Mitylene he will go through port and by land.

48
Seven hundred captives bound roughly.
Lots drawn for the half to be murdered:
The hope at hand will come very promptly
But not as soon as the fifteenth death.

49
Gallic realm, you will be much changed:
To a foreign place is the empire transferred:
You will be set up amidst other customs and laws:
Rouen and Chartres will do much of the worst to you.

50
The republic of the great city
Will not want to consent to the great severity:
King summoned by trumpet to go out,
The ladder at the wall, the city will repent.

51
Paris conspires to commit a great murder
Blois will cause it to be fully carried out:
Those of Orleans will want to replace their chief,
Angers, Troyes, Langres will commit a misdeed against them.

52
In Campania there will be a very long rain,
In Apulia very great drought.
The Cock will see the Eagle, its wing poorly finished,
By the Lion will it be put into extremity.

53
When the greatest one will carry off the prize
Of Nuremberg, of Augsburg, and those of B‰ole
Through Cologne the chief Frankfort retaken
They will cross through Flanders right into Gaul.

54
One of the greatest ones will flee to Spain
Which will thereafter come to bleed in a long wound:
Armies passing over the high mountains,
Devastating all, and then to reign in peace.

55
In the year that one eye will reign in France,
The court will be in very unpleasant trouble:
The great one of Blois will kill his friend:
The realm placed in harm and double doubt.

56
Montauban, N"mes, Avignon and Beziers,
Plague, thunder and hail in the wake of Mars:
Of Paris bridge, Lyons wall, Montpellier,
After six hundreds and seven score three pairs.

57
Seven times will you see the British nation change,
Steeped in blood in 290 years:
Free not at all its support Germanic.
Aries doubt his Bastarnian pole.

58
Near the Rhine from the Noric mountains
Will be born a great one of people come too late,
One who will defend Sarmatia and the Pannonians,
One will not know what will have become of him.

59
Barbarian empire usurped by the third,
The greater part of his blood he will put to death:
Through senile death the fourth struck by him,
For fear that the blood through the blood be not dead.

60
Throughout all Asia (Minor) great proscription,
Even in Mysia, Lycia and Pamphilia.
Blood will be shed because of the absolution
Of a young black one filled with felony.

61
The great band and sect of crusaders
Will be arrayed in Mesopotamia:

Light company of the nearby river,
That such law will hold for an enemy.

62
Near the Douro by the closed Tyrian sea,
He will come to pierce the great Pyrenees mountains.
One hand shorter his opening glosses,
He will lead his traces to Carcassone.

63
The Roman power will be thoroughly abased,
Following in the footsteps of its great neighbor:
Hidden civil hatreds and debates
Will delay their follies for the buffoons.

64
The chief of Persia will occupy great Olchades,
The trireme fleet against the Mahometan people
From Parthia, and Media: and the Cyclades pillaged:
Long rest at the great Ionian port.

65
When the sepulcher of the great Roman is found,
The day after a Pontiff will be elected:
Scarcely will he be approved by the Senate
Poisoned, his blood in the sacred chalice.

66
The great Bailiff of Orleans put to death
Will be by one of blood revengeful:
Of death deserved he will not die, nor by chance:
He made captive poorly by his feet and hands.

67
A new sect of Philosophers
Despising death, gold, honors and riches
Will not be bordering upon the German mountains:
To follow them they will have power and crowds.

68
Leaderless people of Spain and Italy
Dead, overcome within the Peninsula:
Their dictator betrayed by irresponsible folly,
Swimming in blood everywhere in the latitude.

69
The great army led by a young man,
It will come to surrender itself into the hands of the enemies:
But the old one born to the half-pig,
He will cause Ch‰olon and M‰ocon to be friends.

70
The great Britain including England
Will come to be flooded very high by waters
The new League of Ausonia will make war,
So that they will come to strive against them.

71
Those in the isles long besieged
Will take vigor and force against their enemies:
Those outside dead overcome by hunger,
They will be put in greater hunger than ever before.

72
The good old man buried quite alive,
Near the great river through false suspicion:
The new old man ennobled by riches,
Captured on the road all his gold for ransom.

73
When the cripple will attain to the realm,
For his competitor he will have a near bastard:
He and the realm will become so very mangy
That before he recovers, it will be too late.

74
Naples, Florence, Faenza and Imola,
They will be on terms of such disagreement
As to delight in the wretches of Nola
Complaining of having mocked its chief.

75
Pau, Verona, Vicenza, Saragossa,
From distant swords lands wet with blood:
Very great plague will come with the great shell,
Relief near, and the remedies very far.

76
In Germany will be born diverse sects,
Coming very near happy paganism,
The heart captive and returns small,
They will return to paying the true tithe.

77
The third climate included under Aries
The year 1727 in October,
The King of Persia captured by those of Egypt:
Conflict, death, loss: to the cross great shame.

78
Captive of the Eastern seamen:
They will pass Gibraltar and Spain,

Present in Persia for the fearful new King.

79
The fatal everlasting order through the chain
Will come to turn through consistent order:
The chain of Marseilles will be broken:
The city taken, the enemy at the same time.

80
The worthy one chased out of the English realm,
The adviser through anger put to the fire:
His adherents will go so low to efface themselves
That the bastard will be half received.

81
The great shameless, audacious bawler,
He will be elected governor of the army:
The boldness of his contention,
The bridge broken, the city faint from fear.

82
Frejus, Antibes, towns around Nice,
They will be thoroughly devastated by sea and by land:
The locusts by land and by sea the wind propitious,
Captured, dead, bound, pillaged without law of war.

83
The long hairs of Celtic Gaul
Accompanied by foreign nations,
They will make captive the people of Aquitaine,
For succumbing to their designs.

84
The great city will be thoroughly desolated,
Of the inhabitants not a single one will remain there:
Wall, sex, temple and virgin violated,
Through sword, fire, plague, cannon people will die.

85
The city taken through deceit and guile,
Taken in by means of a handsome youth:
Assault given by the Robine near the Aude,
He and all dead for having thoroughly deceived.

86
A chief of Ausonia will go to Spain
By sea, he will make a stop in Marseilles:
Before his death he will linger a long time:
After his death one will see a great marvel.

87
Gallic fleet, do not approach Corsica,
Less Sardinia, you will rue it:
Every one of you will die frustrated of the help of the cape:
You will swim in blood, captive you will not believe me.

88
From Barcelona a very great army by sea,
All Marseilles will tremble with terror:
Isles seized help shut off by sea,
Your traitor will swim on land.

89
At that time Cyprus will be frustrated
Of its relief by those of the Aegean Sea:
Old ones slaughtered: but by speeches and supplications
Their King seduced, Queen outraged more.

90
The great Satyr and Tiger of Hyrcania,
Gift presented to those of the Ocean:
A fleet's chief will set out from Carmania,
One who will take land at the Tyrren Phocaean.

91
The tree which had long been dead and withered,
In one night it will come to grow green again:
The Cronian King sick, Prince with club foot,
Feared by his enemies he will make his sail bound.

92
The world near the last period,
Saturn will come back again late:
Empire transferred towards the Dusky nation,
The eye plucked out by the Goshawk at Narbonne.

93
In Avignon the chief of the whole empire
Will make a stop on the way to desolated Paris:
Tricast will hold the anger of Hannibal:
Lyons will be poorly consoled for the change.

94
For five hundred years more one will keep count of him
Who was the ornament of his time:
Then suddenly great light will he give,
He who for this century will render them very satisfied.

95
The law of More will be seen to decline:
After another much more seductive:

Dnieper first will come to give way:
Through gifts and tongue another more attractive.

96
The Chief of Fossano will have his throat cut
By the leader of the bloodhound and greyhound:
The deed executed by those of the Tarpeian Rock,
Saturn in Leo February 13.

97
New law to occupy the new land
Towards Syria, Judea and Palestine:
The great barbarian empire to decay,
Before the Moon completes it cycle.

98
Two royal brothers will wage war so fiercely
That between them the war will be so mortal
That both will occupy the strong places:
Their great quarrel will fill realm and life.

99
In the grassy fields of Alleins and Vern gues
Of the Luberon range near the Durance,
The conflict will be very sharp for both armies,
Mesopotamia will fail in France.

100
The last one honored amongst the Gauls,
Over the enemy man will he be victorious:
Force and land in a moment explored,
When the envious one will die from an arrow shot.

Century IV

1
That of the remainder of blood unshed:
Venice demands that relief be given:
After having waited a very long time,
City delivered up at the first sound of the horn.

2
Because of death France will take to making a journey,
Fleet by sea, marching over the Pyrenees Mountains,
Spain in trouble, military people marching:
Some of the greatest Ladies carried off to France.

3
From Arras and Bourges many banners of Dusky Ones,
A greater number of Gascons to fight on foot,

Those along the Rhine will bleed the Spanish:
Near the mountain where Sagunto sits.

4
The impotent Prince angry, complaints and quarrels,
Rape and pillage, by cocks and Africans:
Great it is by land, by sea infinite sails,
Italy alone will be chasing Celts.

5
Cross, peace, under one the divine word accomplished,
Spain and Gaul will be united together:
Great disaster near, and combat very bitter:
No heart will be so hardy as not to tremble.

6
By the new clothes after the find is made,
Malicious plot and machination:
First will die he who will prove it,
Color Venetian trap.

7
The minor son of the great and hated Prince,
He will have a great touch of leprosy at the age of twenty:
Of grief his mother will die very sad and emaciated,
And he will die where the loose flesh falls.

8
The great city by prompt and sudden assault
Surprised at night, guards interrupted:
The guards and watches of Saint–Quentin
Slaughtered, guards and the portals broken.

9
The chief of the army in the middle of the crowd
Will be wounded by an arrow shot in the thighs,
When Geneva in tears and distress
Will be betrayed by Lausanne and the Swiss.

10
The young Prince falsely accused
Will plunge the army into trouble and quarrels:
The chief murdered for his support,
Scepter to pacify: then to cure scrofula.

11.
He who will have the government of the great cope
Will be prevailed upon to perform several deeds:
The twelve red one who will come to soil the cloth,
Under murder, murder will come to be perpetrated.

12
The greater army put to flight in disorder,
Scarcely further will it be pursued:
Army reassembled and the legion reduced,
Then it will be chased out completely from the Gauls.

13
News of the greater loss reported,
The report will astonish the army:
Troops united against the revolted:
The double phalanx will abandon the great one.

14
The sudden death of the first personage
Will have caused a change and put another in the sovereignty:
Soon, late come so high and of low age,
Such by land and sea that it will be necessary to fear him.

15
From where they will think to make famine come,
From there will come the surfeit:
The eye of the sea through canine greed
For the one the other will give oil and wheat.

16
The city of liberty made servile:
Made the asylum of profligates and dreamers.
The King changed to them not so violent:
From one hundred become more than a thousand.

17
To change at Beaune, Nuits, Châlon and Dijon,
The duke wishing to improve the Carmelite [nun]
Marching near the river, fish, diver's beak
Will see the tail: the gate will be locked.

18
Some of those most lettered in the celestial facts
Will be condemned by illiterate princes:
Punished by Edict, hunted, like criminals,
And put to death wherever they will be found.

19
Before Rouen the siege laid by the Insubrians,
By land and sea the passages shut up:
By Hainaut and Flanders, by Ghent and those of Liege
Through cloaked gifts they will ravage the shores.

20
Peace and plenty for a long time the place will praise:
Throughout his realm the fleur-de-lis deserted:

Bodies dead by water, land one will bring there,
Vainly awaiting the good fortune to be buried there.

21
The change will be very difficult:
City and province will gain by the change:
Heart high, prudent established, chased out one cunning,
Sea, land, people will change their state.

22
The great army will be chased out,
In one moment it will be needed by the King:
The faith promised from afar will be broken,
He will be seen naked in pitiful disorder.

23
The legion in the marine fleet
Will burn lime, lodestone sulfur and pitch:
The long rest in the secure place:
Port Selyn and Monaco, fire will consume them.

24
Beneath the holy earth of a soul the faint voice heard,
Human flame seen to shine as divine:
It will cause the earth to be stained with the blood of the monks,
And to destroy the holy temples for the impure ones.

25
Lofty bodies endlessly visible to the eye,
Through these reasons they will come to obscure:
Body, forehead included, sense and head invisible,
Diminishing the sacred prayers.

26
The great swarm of bees will arise,
Such that one will not know whence they have come;
By night the ambush, the sentinel under the vines
City delivered by five babblers not naked.

27
Salon, Tarascon, Mausol, the arch of SEX.,
Where the pyramid is still standing:
They will come to deliver the Prince of Annemark,
Redemption reviled in the temple of Artemis.

28
When Venus will be covered by the Sun,
Under the splendor will be a hidden form:
Mercury will have exposed them to the fire,
Through warlike noise it will be insulted.

29
The Sun hidden eclipsed by Mercury
Will be placed only second in the sky:
Of Vulcan Hermes will be made into food,
The Sun will be seen pure, glowing red and golden.

30
Eleven more times the Moon the Sun will not want,
All raised and lowered by degree:
And put so low that one will stitch little gold:
Such that after famine plague, the secret uncovered.

31
The Moon in the full of night over the high mountain,
The new sage with a lone brain sees it:
By his disciples invited to be immortal,
Eyes to the south. Hands in bosoms, bodies in the fire.

32
In the places and times of flesh giving way to fish,
The communal law will be made in opposition:
It will hold strongly the old ones, then removed from the midst,
Loving of Everything in Common put far behind.

33
Jupiter joined more to Venus than to the Moon
Appearing with white fullness:
Venus hidden under the whiteness of Neptune
Struck by Mars through the white stew.

34
The great one of the foreign land led captive,
Chained in gold offered to King Chyren:
He who in Ausonia, Milan will lose the war,
And all his army put to fire and sword.

35
The fire put out the virgins will betray
The greater part of the new band:
Lightning in sword and lance the lone Kings will guard
Etruria and Corsica, by night throat cut.

36
The new sports set up again in Gaul,
After victory in the Insubrian campaign:
Mountains of Hesperia, the great ones tied and trussed up:
Romania and Spain to tremble with fear.

37
The Gaul will come to penetrate the mountains by leaps:
He will occupy the great place of Insubria:

His army to enter to the greatest depth,
Genoa and Monaco will drive back the red fleet.

38
While he will engross the Duke, King and Queen
With the captive Byzantine chief in Samothrace:
Before the assault one will eat the order:
Reverse side metaled will follow the trail of the blood.

39
The Rhodians will demand relief,
Through the neglect of its heirs abandoned.
The Arab empire will reveal its course,
The cause set right again by Hesperia.

40
The fortresses of the besieged shut up,
Through gunpowder sunk into the abyss:
The traitors will all be stowed away alive,
Never did such a pitiful schism happen to the sextons.

41
Female sex captive as a hostage
Will come by night to deceive the guards:
The chief of the army deceived by her language
Will abandon her to the people, it will be pitiful to see.

42
Geneva and Langres through those of Chartres and Dile
And through Grenoble captive at Montelimar
Seyssel, Lausanne, through fraudulent deceit,
They will betray them for sixty marks of gold.

43
Arms will be heard clashing in the sky:
That very same year the divine ones enemies:
They will want unjustly to discuss the holy laws:
Through lightning and war the complacent one put to death.

44
Two large ones of Mende, of Rodez and Milhau
Cahors, Limoges, Castres bad week
By night the entry, from Bordeaux an insult
Through Perigord at the peal of the bell.

45
Through conflict a King will abandon his realm:
The greatest chief will fail in time of need:
Dead, ruined few will escape it,
All cut up, one will be a witness to it.

46
The fact well defended by excellence,
Guard yourself Tours from your near ruin:
London and Nantes will make a defense through Reims
Not passing further in the time of the drizzle.

47
The savage black one when he will have tried
His bloody hand at fire, sword and drawn bows:
All of his people will be terribly frightened,
Seeing the greatest ones hung by neck and feet.

48
The fertile, spacious Ausonian plain
Will produce so many gadflies and locusts,
The solar brightness will become clouded,
All devoured, great plague to come from them.

49
Before the people blood will be shed,
Only from the high heavens will it come far:
But for a long time of one nothing will be heard,
The spirit of a lone one will come to bear witness against it.

50
Libra will see the Hesperias govern,
Holding the monarchy of heaven and earth:
No one will see the forces of Asia perished,
Only seven hold the hierarchy in order.

51
A Duke eager to follow his enemy
Will enter within impeding the phalanx:
Hurried on foot they will come to pursue so closely
That the day will see a conflict near Ganges.

52
In the besieged city men and woman to the walls,
Enemies outside the chief ready to surrender:
The wind will be strongly against the troops,
They will be driven away through lime, dust and ashes.

53
The fugitives and exiles recalled:
Fathers and sons great garnishing of the deep wells:
The cruel father and his people choked:
His far worse son submerged in the well.

54
Of the name which no Gallic King ever had
Never was there so fearful a thunderbolt,

Italy, Spain and the English trembling,
Very attentive to a woman and foreigners.

55
When the crow on the tower made of brick
For seven hours will continue to scream:
Death foretold, the statue stained with blood,
Tyrant murdered, people praying to their Gods.

56
After the victory of the raving tongue,
The spirit tempered in tranquillity and repose:
Throughout the conflict the bloody victor makes orations,
Roasting the tongue and the flesh and the bones.

57
Ignorant envy upheld before the great King,
He will propose forbidding the writings:
His wife not his wife tempted by another,
Twice two more neither skill nor cries.

58
To swallow the burning Sun in the throat,
The Etruscan land washed by human blood:
The chief pail of water, to lead his son away,
Captive lady conducted into Turkish land.

59
Two beset in burning fervor:
By thirst for two full cups extinguished,
The fort filed, and an old dreamer,
To the Genevans he will show the track from Nira.

60
The seven children left in hostage,
The third will come to slaughter his child:
Because of his son two will be pierced by the point,
Genoa, Florence, he will come to confuse them.

61
The old one mocked and deprived of his place,
By the foreigner who will suborn him:
Hands of his son eaten before his face,
His brother to Chartres, Orleans Rouen will betray.

62
A colonel with ambition plots,
He will seize the greatest army,
Against his Prince false invention,
And he will be discovered under his arbor.

63
The Celtic army against the mountaineers,
Those who will be learned and able in bird-calling:
Peasants will soon work fresh presses,
All hurled on the sword's edge.

64
The transgressor in bourgeois garb,
He will come to try the King with his offense:
Fifteen soldiers for the most part bandits,
Last of life and chief of his fortune.

65
Towards the deserter of the great fortress,
After he will have abandoned his place,
His adversary will exhibit very great prowess,
The Emperor soon dead will be condemned.

66
Under the feigned color of seven shaven heads
Diverse spies will be scattered:
Wells and fountains sprinkled with poisons,
At the fort of Genoa devourers of men.

67
The year that Saturn and Mars are equal fiery,
The air very dry parched long meteor:
Through secret fires a great place blazing from burning heat,
Little rain, warm wind, wars, incursions.

68
The two greatest ones of Asia and of Africa,
From the Rhine and Lower Danube they will be said to have come,
Cries, tears at Malta and the Ligurian side.

69
The exiles will hold the great city,
The citizens dead, murdered and driven out:
Those of Aquileia will promise Parma
To show them the entry through the untracked places.

70
Quite contiguous to the great Pyrenees mountains,
One to direct a great army against the Eagle:
Veins opened, forces exterminated,
As far as Pau will he come to chase the chief.

71
In place of the bride the daughters slaughtered,
Murder with great error no survivor to be:
Within the well vestals inundated,

The bride extinguished by a drink of Aconite.

72
Those of N"mes through Agen and Lectoure
At Saint–Felix will hold their parliament:
Those of Bazas will come at the unhappy hour
To seize Condom and Marsan promptly.

73
The great nephew by force will test
The treaty made by the pusillanimous heart:
The Duke will try Ferrara and Asti,
When the pantomime will take place in the evening.

74
Those of lake Geneva and of M‰con:
All assembled against those of Aquitaine:
Many Germans many more Swiss,
They will be routed along with those of the Humane.

75
Ready to fight one will desert,
The chief adversary will obtain the victory:
The rear guard will make a defense,
The faltering ones dead in the white territory.

76
Will be vexed, holding as far as the Rhine:
The union of Gascons and Bigorre
To betray the temple, the priest giving his sermon.

77
Selin monarch Italy peaceful,
Realms united by the Christian King of the World:
Dying he will want to lie in Blois soil,
After having chased the pirates from the sea.

78
The great army of the civil struggle,
By night Parma to the foreign one discovered,
Seventy–nine murdered in the town,
The foreigners all put to the sword.

79
Blood Royal flee, Monheurt, Mas, Aiguillon,
The Landes will be filled by Bordelais,
Navarre, Bigorre points and spurs,
Deep in hunger to devour acorns of the cork oak.

80
Near the great river, great ditch, earth drawn out,

In fifteen parts will the water be divided:
The city taken, fire, blood, cries, sad conflict,
And the greatest part involving the coliseum.

81
Promptly will one build a bridge of boats,
To pass the army of the great Belgian Prince:
Poured forth inside and not far from Brussels,
Passed beyond, seven cut up by pike.

82
A throng approaches coming from Slavonia,
The old Destroyer the city will ruin:
He will see his Romania quite desolated,
Then he will not know how to put out the great flame.

83
Combat by night the valiant captain
Conquered will flee few people conquered:
His people stirred up, sedition not in vain,
His own son will hold him besieged.

84
A great one of Auxerre will die very miserable,
Driven out by those who had been under him:
Put in chains, behind a strong cable,
In the year that Mars, Venus and Sun are in conjunction in summer.

85
The white coal will be chased by the black one,
Made prisoner led to the dung cart,
Moor Camel on twisted feet,
Then the younger one will blind the hobby falcon.

86
The year that Saturn will be conjoined in Aquarius
With the Sun, the very powerful King
Will be received and anointed at Reims and Aix,
After conquests he will murder the innocent.

87
A King's son learned in many languages,
Different from his senior in the realm:
His handsome father understood by the greater son,
He will cause his principal adherent to perish.

88
Anthony by name great by the filthy fact
Of Lousiness wasted to his end:
One who will want to be desirous of lead,
Passing the port he will be immersed by the elected one.

89
Thirty of London will conspire secretly
Against their King, the enterprise on the bridge:
He and his satellites will have a distaste for death,
A fair King elected, native of Frisia.

90
The two armies will be unable to unite at the walls,
In that instant Milan and Pavia to tremble:
Hunger, thirst, doubt will come to plague them very strongly
They will not have a single morsel of meat, bread or victuals.

91
For the Gallic Duke compelled to fight in the duel,
The ship of Melilla will not approach Monaco,
Wrongly accused, perpetual prison,
His son will strive to reign before his death.

92
The head of the valiant captain cut off,
It will be thrown before his adversary:
His body hung on the sail-yard of the ship,
Confused it will flee by oars against the wind.

93
A serpent seen near the royal bed,
It will be by the lady at night the dogs will not bark:
Then to be born in France a Prince so royal,
Come from heaven all the Princes will see him.

94
Two great brothers will be chased out of Spain,
The elder conquered under the Pyrenees mountains:
The sea to redden, Rhine, bloody Lake Geneva from Germany,
Narbonne, Beziers contaminated by Agde.

95
The realm left to two they will hold it very briefly,
Three years and seven months passed by they will make war:
The two Vestals will rebel in opposition,
Victor the younger in the land of Brittany.

96
The elder sister of the British Isle
Will be born fifteen years before her brother,
Because of her promise procuring verification,
She will succeed to the kingdom of the balance.

97
The year that Mercury, Mars, Venus in retrogression,
The line of the great Monarch will not fail:

Elected by the Portuguese people near Cadiz,
One who will come to grow very old in peace and reign.

98
Those of Alba will pass into Rome,
By means of Langres the multitude muffled up,
Marquis and Duke will pardon no man,
Fire, blood, smallpox no water the crops to fail.

99
The valiant elder son of the King's daughter,
He will hurl back the Celts very far,
Such that he will cast thunderbolts, so many in such an array
Few and distant, then deep into the Hesperias.

100
From the celestial fire on the Royal edifice,
When the light of Mars will go out,
Seven months great war, people dead through evil
Rouen, Evreux the King will not fail.

Century V

1
Before the coming of Celtic ruin,
In the temple two will parley
Pike and dagger to the heart of one mounted on the steed,
They will bury the great one without making any noise.

2
Seven conspirators at the banquet will cause to flash
The iron out of the ship against the three:
One will have the two fleets brought to the great one,
When through the evil the latter shoots him in the forehead.

3
The successor to the Duchy will come,
Very far beyond the Tuscan Sea:
A Gallic branch will hold Florence,
The nautical Frog in its bosom be agreement.

4
The large mastiff expelled from the city
Will be vexed by the strange alliance,
After having chased the stag to the fields
The wolf and the Bear will defy each other.

5
Under the shadowy pretense of removing servitude,
He will himself usurp the people and city:

He will do worse because of the deceit of the young prostitute,
Delivered in the field reading the false poem.

6
The Augur putting his hand upon the head of the King
Will come to pray for the peace of Italy:
He will come to move the scepter to his left hand,
From King he will become pacific Emperor.

7
The bones of the Triumvir will be found,
Looking for a deep enigmatic treasure:
Those from thereabouts will not be at rest,
Digging for this thing of marble and metallic lead.

8
There will be unleashed live fire, hidden death,
Horrible and frightful within the globes,
By night the city reduced to dust by the fleet,
The city afire, the enemy amenable.

9
The great arch demolished down to its base,
By the chief captive his friend forestalled,
He will be born of the lady with hairy forehead and face,
Then through cunning the Duke overtaken by death.

10
A Celtic chief wounded in the conflict
Seeing death overtaking his men near a cellar:
Pressed by blood and wounds and enemies,
And relief by four unknown ones.

11
The sea will not be passed over safely by those of the Sun,
Those of Venus will hold all Africa:
Saturn will no longer occupy their realm,
And the Asiatic part will change.

12
To near the Lake of Geneva will it be conducted,
By the foreign maiden wishing to betray the city:
Before its murder at Augsburg the great suite,
And those of the Rhine will come to invade it.

13
With great fury the Roman Belgian King
Will want to vex the barbarian with his phalanx:
Fury gnashing, he will chase the African people
From the Pannonias to the pillars of Hercules.

14
Saturn and Mars in Leo Spain captive,
By the African chief trapped in the conflict,
Near Malta, Herod taken alive,
And the Roman scepter will be struck down by the Cock.

15
The great Pontiff taken captive while navigating,
The great one thereafter to fail the clergy in tumult:
Second one elected absent his estate declines,
His favorite bastard to death broken on the wheel.

16
The Sabaean tear no longer at its high price,
Turning human flesh into ashes through death,
At the isle of Pharos disturbed by the Crusaders,
When at Rhodes will appear a hard phantom.

17
By night the King passing near an Alley,
He of Cyprus and the principal guard:
The King mistaken, the hand flees the length of the Rhine,
The conspirators will set out to put him to death.

18
The unhappy abandoned one will die of grief,
His conqueress will celebrate the hecatomb:
Pristine law, free edict drawn up,
The wall and the Prince falls on the seventh day.

19
The great Royal one of gold, augmented by brass,
The agreement broken, war opened by a young man:
People afflicted because of a lamented chief,
The land will be covered with barbarian blood.

20
The great army will pass beyond the Alps,
Shortly before will be born a monster scoundrel:
Prodigious and sudden he will turn
The great Tuscan to his nearest place.

21
By the death of the Latin Monarch,
Those whom he will have assisted through his reign:
The fire will light up again the booty divided,
Public death for the bold ones who incurred it.

22
Before the great one has given up the ghost at Rome,
Great terror for the foreign army:

The ambush by squadrons near Parma,
Then the two red ones will celebrate together.

23
The two contented ones will be united together,
When for the most part they will be conjoined with Mars:
The great one of Africa trembles in terror,
Duumvirate disjoined by the fleet.

24
The realm and law raised under Venus,
Saturn will have dominion over Jupiter:
The law and realm raised by the Sun,
Through those of Saturn it will suffer the worst.

25
The Arab Prince Mars, Sun, Venus, Leo,
The rule of the Church will succumb by sea:
Towards Persia very nearly a million men,
The true serpent will invade Byzantium and Egypt.

26
The slavish people through luck in war
Will become elevated to a very high degree:
They will change their Prince, one born a provincial,
An army raised in the mountains to pass over the sea.

27
Through fire and arms not far from the Black Sea,
He will come from Persia to occupy Trebizond:
Pharos, Mytilene to tremble, the Sun joyful,
The Adriatic Sea covered with Arab blood.

28
His arm hung and leg bound,
Face pale, dagger hidden in his bosom,
Three who will be sworn in the fray
Against the great one of Genoa will the steel be unleashed.

29
Liberty will not be recovered,
A proud, villainous, wicked black one will occupy it,
When the matter of the bridge will be opened,
The republic of Venice vexed by the Danube.

30
All around the great city
Soldiers will be lodged throughout the fields and towns:
To give the assault Paris, Rome incited,
Then upon the bridge great pillage will be carried out.

31
Through the Attic land fountain of wisdom,
At present the rose of the world:
The bridge ruined, and its great pre-eminence
Will be subjected, a wreck amidst the waves.

32
Where all is good, the Sun all beneficial and the Moon
Is abundant, its ruin approaches:
From the sky it advances to change your fortune.
In the same state as the seventh rock.

33
Of the principal ones of the city in rebellion
Who will strive mightily to recover their liberty:
The males cut up, unhappy fray,
Cries, groans at Nantes pitiful to see.

34
From the deepest part of the English West
Where the head of the British isle is
A fleet will enter the Gironde through Blois,
Through wine and salt, fires hidden in the casks.

35
For the free city of the great Crescent sea,
Which still carries the stone in its stomach,
The English fleet will come under the drizzle
To seize a branch, war opened by the great one.

36
The sister's brother through the quarrel and deceit
Will come to mix dew in the mineral:
On the cake given to the slow old woman,
She dies tasting it she will be simple and rustic.

37
Three hundred will be in accord with one will
To come to the execution of their blow,
Twenty months after all memory
Their king betrayed simulating feigned hate.

38
He who will succeed the great monarch on his death
Will lead an illicit and wanton life:
Through nonchalance he will give way to all,
So that in the end the Salic law will fail.

39
Issued from the true branch of the fleur-de-lis,
Placed and lodged as heir of Etruria:

His ancient blood woven by long hand,
He will cause the escutcheon of Florence to bloom.

40
The blood royal will be so very mixed,
Gauls will be constrained by Hesperia:
One will wait until his term has expired,
And until the memory of his voice has perished.

41
Born in the shadows and during a dark day,
He will be sovereign in realm and goodness:
He will cause his blood to rise again in the ancient urn,
Renewing the age of gold for that of brass.

42
Mars raised to his highest belfry
Will cause the Savoyards to withdraw from France:
The Lombard people will cause very great terror
To those of the Eagle included under the Balance.

43
The great ruin of the holy things is not far off,
Provence, Naples, Sicily, Sees and Pons:
In Germany, at the Rhine and Cologne,
Vexed to death by all those of Mainz.

44
On sea the red one will be taken by pirates,
Because of him peace will be troubled:
Anger and greed will he expose through a false act,
The army doubled by the great Pontiff.

45
The great Empire will soon be desolated
And transferred to near the Ardennes:
The two bastards beheaded by the oldest one,
And Bronzebeard the hawk–nose will reign.

46
Quarrels and new schism by the red hats
When the Sabine will have been elected:
They will produce great sophism against him,
And Rome will be injured by those of Alba.

47
The great Arab will march far forward,
He will be betrayed by the Byzantians:
Ancient Rhodes will come to meet him,
And greater harm through the Austrian Hungarians.

48
After the great affliction of the scepter,
Two enemies will be defeated by them:
A fleet from Africa will appear before the Hungarians,
By land and sea horrible deeds will take place.

49
Not from Spain but from ancient France
Will one be elected for the trembling bark,
To the enemy will a promise be made,
He who will cause a cruel plague in his realm.

50
The year that the brothers of the lily come of age,
One of them will hold the great Romania:
The mountains to tremble, Latin passage opened,
Agreement to march against the fort of Armenia.

51
The people of Dacia, England, Poland
And of Bohemia will make a new league:
To pass beyond the pillars of Hercules,
The Barcelonians and Tuscans will prepare a cruel plot.

52
There will be a King who will give opposition,
The exiles raised over the realm:
The pure poor people to swim in blood,
And for a long time will he flourish under such a device.

53
The law of the Sun and of Venus in strife,
Appropriating the spirit of prophecy:
Neither the one nor the other will be understood,
The law of the great Messiah will hold through the Sun.

54
From beyond the Black Sea and great Tartary,
There will be a King who will come to see Gaul,
He will pierce through Alania and Armenia,
And within Byzantium will he leave his bloody rod.

55
In the country of Arabia Felix
There will be born one powerful in the law of Mahomet:
To vex Spain, to conquer Grenada,
And more by sea against the Ligurian people.

56
Through the death of the very old Pontiff
A Roman of good age will be elected,

Of him it will be said that he weakens his see,
But long will he sit and in biting activity.

57
There will go from Mont and Aventin,
One who through the hole will warn the army:
Between two rocks will the booty be taken,
Of Sectus' mausoleum the renown to fail.

58
By the aqueduct of Uz s over the Gard,
Through the forest and inaccessible mountain,
In the middle of the bridge there will be cut in the fist
The chief of N"mes who will be very terrible.

59
Too long a stay for the English chief at N"mes,
Towards Spain Redbeard to the rescue:
Many will die by war opened that day,
When a bearded star will fall in Artois.

60
By the shaven head a very bad choice will come to be made,
Overburdened he will not pass the gate:
He will speak with such great fury and rage,
That to fire and blood he will consign the entire sex.

61
The child of the great one not by his birth,
He will subjugate the high Apennine mountains:
He will cause all those of the balance to tremble,
And from the Pyrenees to Mont Cenis.

62
One will see blood to rain on the rocks,
Sun in the East, Saturn in the West:
Near Orgon war, at Rome great evil to be seen,
Ships sunk to the bottom, taken by Trident.

63
From the vain enterprise honor and undue complaint,
Boats tossed about among the Latins, cold, hunger, waves
Not far from the Tiber the land stained with blood,
And diverse plagues will be upon mankind.

64
Those assembled by the tranquillity of the great number,
By land and sea counsel countermanded:
Near Antonne Genoa, Nice in the shadow
Through fields and towns in revolt against the chief.

65
Come suddenly the terror will be great,
Hidden by the principal ones of the affair:
And the lady on the charcoal will no longer be in sight,
Thus little by little will the great ones be angered.

66
Under the ancient vestal edifices,
Not far from the ruined aqueduct:
The glittering metals are of the Sun and Moon,
The lamp of Trajan engraved with gold burning.

67
When the chief of Perugia will not venture his tunic
Sense under cover to strip himself quite naked:
Seven will be taken Aristocratic deed,
Father and son dead through a point in the collar.

68
In the Danube and of the Rhine will come to drink
The great Camel, not repenting it:
Those of the Rhine to tremble, and much more so those of the Loire,
and near the Alps the Cock will ruin him.

69
No longer will the great one be in his false sleep,
Uneasiness will come to replace tranquillity:
A phalanx of gold, azure and vermilion arrayed
To subjugate Africa and gnaw it to the bone,

70
Of the regions subject to the Balance,
They will trouble the mountains with great war,
Captives the entire sex enthralled and all Byzantium,
So that at dawn they will spread the news from land to land.

71
By the fury of one who will wait for the water,
By his great rage the entire army moved:
Seventeen boats loaded with the noble,
The messenger come late along the Rhine.

72
For the pleasure of the voluptuous edict,
One will mix poison in the faith:
Venus will be in a course so virtuous
As to becloud the whole quality of the Sun.

73
The Church of God will be persecuted,
And the holy Temples will be plundered,

The child will put his mother out in her shift,
Arabs will be allied with the Poles.

74
Of Trojan blood will be born a Germanic heart
Who will rise to very high power:
He will drive out the foreign Arabic people,
Returning the Church to its pristine pre-eminence.

75
He will rise high over the estate more to the right,
He will remain seated on the square stone,
Towards the south facing to his left,
The crooked staff in his hand his mouth sealed.

76
In a free place will he pitch his tent,
And he will not want to lodge in the cities:
Aix, Carpentras, L'Isle, Vaucluse Mont, Cavaillon,
Throughout all these places will he abolish his trace.

77
All degrees of Ecclesiastical honor
Will be changed to that of Jupiter and Quirinus:
The priest of Quirinus to one of Mars,
Then a King of France will make him one of Vulcan.

78
The two will not be united for very long,
And in thirteen years to the Barbarian Satrap:
On both sides they will cause such loss
That one will bless the Bark and its cope.

79
The sacred pomp will come to lower its wings,
Through the coming of the great legislator:
He will raise the humble, he will vex the rebels,
His like will not appear on this earth.

80
Ogmios will approach great Byzantium,
The Barbaric League will be driven out:
Of the two laws the heathen one will give way,
Barbarian and Frank in perpetual strife.

81
The royal bird over the city of the Sun,
Seven months in advance it will deliver a nocturnal omen:
The Eastern wall will fall lightning thunder,
Seven days the enemies directly to the gates.

82
At the conclusion of the treaty outside the fortress
Will not go he who is placed in despair:
When those of Arbois, of Langres against Bresse
Will have the mountains of Dôle an enemy ambush.

83
Those who will have undertaken to subvert,
An unparalleled realm, powerful and invincible:
They will act through deceit, nights three to warn,
When the greatest one will read his Bible at the table.

84
He will be born of the gulf and unmeasured city,
Born of obscure and dark family:
He who the revered power of the great King
Will want to destroy through Rouen and Evreux.

85
Through the Suevi and neighboring places,
They will be at war over the clouds:
Swarm of marine locusts and gnats,
The faults of Geneva will be laid quite bare.

86
Divided by the two heads and three arms,
The great city will be vexed by waters:
Some great ones among them led astray in exile,
Byzantium hard pressed by the head of Persia.

87
The year that Saturn is out of bondage,
In the Frank land he will be inundated by water:
Of Trojan blood will his marriage be,
And he will be confined safely be the Spaniards.

88
Through a frightful flood upon the sand,
A marine monster from other seas found:
Near the place will be made a refuge,
Holding Savona the slave of Turin.

89
Into Hungary through Bohemia, Navarre,
and under that banner holy insurrections:
By the fleur–de–lis legion carrying the bar,
Against Orleans they will cause disturbances.

90
In the Cyclades, in Perinthus and Larissa,
In Sparta and the entire Pelopennesus:

Very great famine, plague through false dust,
Nine months will it last and throughout the entire peninsula.

91
At the market that they call that of liars,
Of the entire Torrent and field of Athens:
They will be surprised by the light horses,
By those of Alba when Mars is in Leo and Saturn in Aquarius.

92
After the see has been held seventeen years,
Five will change within the same period of time:
Then one will be elected at the same time,
One who will not be too comfortable to the Romans.

93
Under the land of the round lunar globe,
When Mercury will be dominating:
The isle of Scotland will produce a luminary,
One who will put the English into confusion.

94
He will transfer into great Germany
Brabant and Flanders, Ghent, Bruges and Boulogne:
The truce feigned, the great Duke of Armenia
Will assail Vienna and Cologne.

95
The nautical oar will tempt the shadows,
Then it will come to stir up the great Empire:
In the Aegean Sea the impediments of wood
Obstructing the diverted Tyrrhenian Sea.

96
The rose upon the middle of the great world,
For new deeds public shedding of blood:
To speak the truth, one will have a closed mouth,
Then at the time of need the awaited one will come late.

97
The one born deformed suffocated in horror,
In the habitable city of the great King:
The severe edict of the captives revoked,
Hail and thunder, Condom inestimable.

98
At the forty-eighth climacteric degree,
At the end of Cancer very great dryness:
Fish in sea, river, lake boiled hectic,
Bearn, Bigorre in distress through fire from the sky.

99
Milan, Ferrara, Turin and Aquileia,
Capua, Brindisi vexed by the Celtic nation:
By the Lion and his Eagleâs phalanx,
When the old British chief Rome will have.

100
The incendiary trapped in his own fire,
Of fire from the sky at Carcassonne and the Comminges:
Foix, Auch, Maz res, the high old man escaped,
Through those of Hesse and Thuringia, and some Saxons.

Century VI

1
Around the Pyrenees mountains a great throng
Of foreign people to aid the new King:
Near the great temple of Le Mas by the Garonne,
A Roman chief will fear him in the water.

2
In the year five hundred eighty more or less,
One will await a very strange century:
In the year seven hundred and three the heavens witness thereof,
That several kingdoms one to five will make a change.

3
The river that tries the new Celtic heir
Will be in great discord with the Empire:
The young Prince through the ecclesiastical people
Will remove the scepter of the crown of concord.

4
The Celtic river will change its course,
No longer will it include the city of Agrippina:
All changed except the old language,
Saturn, Leo, Mars, Cancer in plunder.

5
Very great famine through pestiferous wave,
Through long rain the length of the arctic pole:
Samarobryn one hundred leagues from the hemisphere,
The will live without law exempt from politics.

6
There will appear towards the North
Not far from Cancer the bearded star:
Susa, Siena, Boeotia, Eretria,
The great one of Rome will die, the night over.

7
Norway and Dacia and the British Isle
Will be vexed by the united brothers:
The Roman chief sprung from Gallic blood
And his forces hurled back into the forests.

8
Those who were in the realm for knowledge
Will become impoverished at the change of King:
Some exiled without support, having no gold,
The lettered and letters will not be at a high premium.

9
In the sacred temples scandals will be perpetrated,
They will be reckoned as honors and commendations:
Of one of whom they engrave medals of silver and of gold,
The end will be in very strange torments.

10
In a short time the temples with colors
Of white and black of the two intermixed:
Red and yellow ones will carry off theirs from them,
Blood, land, plague, famine, fire extinguished by water.

11
The seven branches will be reduced to three,
The elder ones will be surprised by death,
The two will be seduced to fratricide,
The conspirators will be dead while sleeping.

12
To raise forces to ascend to the empire
In the Vatican the Royal blood will hold fast:
Flemings, English, Spain with Aspire
Against Italy and France will he contend.

13
A doubtful one will not come far from the realm,
The greater part will want to uphold him:
A Capitol will not want him to reign at all,
He will be unable to bear his great burden.

14
Far from his land a King will lose the battle,
At once escaped, pursued, then captured,
Ignorant one taken under the golden mail,
Under false garb, and the enemy surprised.

15
Under the tomb will be found a Prince
Who will be valued above Nuremberg:

The Spanish King in Capricorn thin,
Deceived and betrayed by the great Wittenberg.

16
That which will be carried off by the young Hawk,
By the Normans of France and Picardy:
The black ones of the temple of the Black Forest place
Will make an inn and fire of Lombardy.

17
After the files the ass-drivers burned,
They will be obliged to change diverse garbs:
Those of Saturn burned by the millers,
Except the greater part which will not be covered.

18
The great King abandoned by the Physicians,
By fate not the Jew's art he remains alive,
He and his kindred pushed high in the realm,
Pardon given to the race which denies Christ.

19
The true flame will devour the lady
Who will want to put the Innocent Ones to the fire:
Before the assault the army is inflamed,
When in Seville a monster in beef will be seen.

20
The feigned union will be of short duration,
Some changed most reformed:
In the vessels people will be in suffering,
Then Rome will have a new Leopard.

21
When those of the arctic pole are united together,
Great terror and fear in the East:
Newly elected, the great trembling supported,
Rhodes, Byzantium stained with Barbarian blood.

22
Within the land of the great heavenly temple,
Nephew murdered at London through feigned peace:
The bark will then become schismatic,
Sham liberty will be proclaimed everywhere.

23
Coins depreciated by the spirit of the realm,
And people will be stirred up against their King:
New peace made, holy laws become worse,
Paris was never in so severe an array.

24
Mars and the scepter will be found conjoined
Under Cancer calamitous war:
Shortly afterwards a new King will be anointed,
One who for a long time will pacify the earth.

25
Through adverse Mars will the monarchy
Of the great fisherman be in ruinous trouble:
The young red black one will seize the hierarchy, .
The traitors will act on a day of drizzle.

26
For four years the see will be held with some little good,
One libidinous in life will succeed to it:
Ravenna, Pisa and Verona will give support,
Longing to elevate the Papal cross.

27
Within the Isles of five rivers to one,
Through the expansion of the great Chyren Selin:
Through the drizzles in the air the fury of one,
Six escaped, hidden bundles of flax.

28
The great Celt will enter Rome,
Leading a throng of the exiled and banished:
The great Pastor will put to death every man
Who was united at the Alps for the cock.

29
The saintly widow hearing the news,
Of her offspring placed in perplexity and trouble:
He who will be instructed to appease the quarrels,
He will pile them up by his pursuit of the shaven heads.

30
Through the appearance of the feigned sanctity,
The siege will be betrayed to the enemies:
In the night when they trusted to sleep in safety,
Near Brabant will march those of Li ge.

31
The King will find that which he desired so much
When the Prelate will be blamed unjustly:
His reply to the Duke will leave him dissatisfied,
He who in Milan will put several to death.

32
Beaten to death by rods for treason,
Captured he will be overcome through his disorder:

Frivolous counsel held out to the great captive,
When Berich will come to bite his nose in fury.

33
His last hand through sanguinary,
He will be unable to protect himself by sea:
Between two rivers he will fear the military hand,
The black and irate one will make him rue it.

34
The device of flying fire
Will come to trouble the great besieged chief:
Within there will be such sedition
That the profligate ones will be in despair.

35
Near the Bear and close to the white wool,
Aries, Taurus, Cancer, Leo, Virgo,
Mars, Jupiter, the Sun will burn a great plain,
Woods and cities letters hidden in the candle.

36
Neither good nor evil through terrestrial battle
Will reach the confines of Perugia,
Pisa to rebel, Florence to see an evil existence,
King by night wounded on a mule with black housing.

37
The ancient work will be finished,
Evil ruin will fall upon the great one from the roof:
Dead they will accuse an innocent one of the deed,
The guilty one hidden in the copse in the drizzle.

38
The enemies of peace to the profligates,
After having conquered Italy:
The bloodthirsty black one, red, will be exposed,
Fire, blood shed, water colored by blood.

39
The child of the realm through the capture of his father
Will be plundered to deliver him:
Near the Lake of Perugia the azure captive,
The hostage troop to become far too drunk.

40
To quench the great thirst the great one of Mainz
Will be deprived of his great dignity:
Those of Cologne will come to complain so loudly
That the great rump will be thrown into the Rhine.

41
The second chief of the realm of Annemark,
Through those of Frisia and of the British Isle,
Will spend more than one hundred thousand marks,
Exploiting in vain the voyage to Italy.

42
To Ogmios will be left the realm
Of the great Selin, who will in fact do more:
Throughout Italy will he extend his banner,
He will be ruled by a prudent deformed one.

43
For a long time will she remain uninhabited,
Around where the Seine and the Marne she comes to water:
Tried by the Thames and warriors,
The guards deceived in trusting in the repulse.

44
By night the Rainbow will appear for Nantes,
By marine arts they will stir up rain:
In the Gulf of Arabia a great fleet will plunge to the bottom,
In Saxony a monster will be born of a bear and a sow.

45
The very learned governor of the realm,
Not wishing to consent to the royal deed:
The fleet at Melilla through contrary wind
Will deliver him to his most disloyal one.

46
A just one will be sent back again into exile,
Through pestilence to the confines of Nonseggle,
His reply to the red one will cause him to be misled,
The King withdrawing to the Frog and the Eagle.

47
The two great ones assembled between two mountains
Will abandon their secret quarrel:
Brussels and Dile overcome by Langres,
To execute their plague at Malines.

48
The too false and seductive sanctity,
Accompanied by an eloquent tongue:
The old city, and Parma too premature,
Florence and Siena they will render more desert.

49
The great Pontiff of the party of Mars
Will subjugate the confines of the Danube:

The cross to pursue, through sword hook or crook,
Captives, gold, jewels more than one hundred thousand rubies.

50
Within the pit will be found the bones,
Incest will be committed by the stepmother:
The state changed, they will demand fame and praise,
And they will have Mars attending as their star.

51
People assembled to see a new spectacle,
Princes and Kings amongst many bystanders,
Pillars walls to fall: but as by a miracle
The King saved and thirty of the ones present.

52
In place of the great one who will be condemned,
Outside the prison, his friend in his place:
The Trojan hope in six months joined, born dead,
The Sun in the urn rivers will be frozen.

53
The great Celtic Prelate suspected by the King,
By night in flight he will leave the realm:
Through a Duke fruitful for his great British King,
Byzantium to Cyprus and Tunis unsuspected.

54
At daybreak at the second crowing of the cock,
Those of Tunis, of Fez and of Bougie,
By the Arabs the King of Morocco captured,
The year sixteen hundred and seven, of the Liturgy.

55
By the appeased Duke in drawing up the contract,
Arabesque sail seen, sudden discovery:
Tripoli, Chios, and those of Trebizond,
Duke captured, the Black Sea and the city a desert.

56
The dreaded army of the Narbonne enemy
Will frighten very greatly the Hesperians:
Perpignan empty through the blind one of Arbon,
Then Barcelona by sea will take up the quarrel.

57
He who was well forward in the realm,
Having a red chief close to the hierarchy,
Harsh and cruel, and he will make himself much feared,
He will succeed to the sacred monarchy.

58
Between the two distant monarchs,
When the clear Sun is lost through Selin:
Great enmity between two indignant ones,
So that liberty is restored to the Isles and Siena.

59
The Lady in fury through rage of adultery,
She will come to conspire not to tell her Prince:
But soon will the blame be made known,
So that seventeen will be put to martyrdom.

60
The Prince outside his Celtic land
Will be betrayed, deceived by the interpreter:
Rouen, La Rochelle through those of Brittany
At the port of Blaye deceived by monk and priest.

61
The great carpet folded will not show
But by halved the greatest part of history:
Driven far out of the realm he will appear harsh,
So that everyone will come to believe in his warlike deed.

62
Too late both the flowers will be lost,
The serpent will not want to act against the law:
The forces of the Leaguers confounded by the French,
Savona, Albenga through Monaco great martyrdom.

63
The lady left alone in the realm
By the unique one extinguished first on the bed of honor:
Seven years will she be weeping in grief,
Then with great good fortune for the realm long life.

64
No peace agreed upon will be kept,
All the subscribers will act with deceit:
In peace and truce, land and sea in protest,
By Barcelona fleet seized with ingenuity.

65
Gray and brown in half-opened war,
By night they will be assaulted and pillaged:
The brown captured will pass through the lock,
His temple opened, two slipped in the plaster.

66
At the foundation of the new sect,
The bones of the great Roman will be found,

A sepulcher covered by marble will appear,
Earth to quake in April poorly buried.

67
Quite another one will attain to the great Empire,
Kindness distant more so happiness:
Ruled by one sprung not far from the brothel,
Realms to decay great bad luck.

68
When the soldiers in a seditious fury
Will cause steel to flash by night against their chief:
The enemy Alba acts with furious hand,
Then to vex Rome and seduce the principal ones.

69
The great pity will occur before long,
Those who gave will be obliged to take:
Naked, starving, withstanding cold and thirst,
To pass over the mountains committing a great scandal.

70
Chief of the world will the great Chyren be,
Plus Ultra behind, loved, feared, dreaded:
His fame and praise will go beyond the heavens,
And with the sole title of Victor will he be quite satisfied.

71
When they will come to give the last rites to the great King
Before he has entirely given up the ghost:
He who will come to grieve over him the least,
Through Lions, Eagles, cross crown sold.

72
Through feigned fury of divine emotion
The wife of the great one will be violated:
The judges wishing to condemn such a doctrine,
She is sacrificed a victim to the ignorant people.

73
In a great city a monk and artisan,
Lodged near the gate and walls,
Secret speaking emptily against Modena,
Betrayed for acting under the guise of nuptials.

74
She chased out will return to the realm,
Her enemies found to be conspirators:
More than ever her time will triumph,
Three and seventy to death very sure.

75
The great Pilot will be commissioned by the King,
To leave the fleet to fill a higher post:
Seven years after he will be in rebellion,
Venice will come to fear the Barbarian army.

76
The ancient city the creation of Antenor,
Being no longer able to bear the tyrant:
The feigned handle in the temple to cut a throat,
The people will come to put his followers to death.

77
Through the fraudulent victory of the deceived,
Two fleets one, German revolt:
The chief murdered and his son in the tent,
Florence and Imola pursued into Romania.

78
To proclaim the victory of the great expanding Selin:
By the Romans will the Eagle be demanded,
Pavia, Milan and Genoa will not consent thereto,
Then by themselves the great Lord claimed.

79
Near the Ticino the inhabitants of the Loire,
Garonne and Saine, the Seine, the Tain and Gironde:
They will erect a promontory beyond the mountains,
Conflict given, Po enlarged, submerged in the wave.

80
From Fez the realm will reach those of Europe,
Their city ablaze and the blade will cut:
The great one of Asia by land and sea with great troop,
So that blues and Pers[ians] the cross will pursue to death.

81
Tears, cries and laments, howls, terror,
Heart inhuman, cruel, black and chilly:
Lake of Geneva the Isles, of Genoa the notables,
Blood to pour out, wheat famine to none mercy.

82
Through the deserts of the free and wild place,
The nephew of the great Pontiff will come to wander:
Felled by seven with a heavy club,
By those who afterwards will occupy the Chalice.

83
He who will have so much honor and flattery
At his entry into Belgian Gaul:

A while after he will act very rudely,
And he will act very warlike against the flower.

84
The Lame One, he who lame could not reign in Sparta,
He will do much through seductive means:
So that by the short and long, he will be accused
Of making his perspective against the King.

85
The great city of Tarsus by the Gauls
Will be destroyed, all of the Turban captives:
Help by sea from the great one of Portugal,
First day of summer Urban's consecration.

86
The great Prelate one day after his dream,
Interpreted opposite to its meaning:
From Gascony a monk will come unexpectedly,
One who will cause the great prelate of Sens to be elected.

87
The election made in Frankfort
Will be voided, Milan will be opposed:
The follower closer will seem so very strong
That he will drive him out into the marshes beyond the Rhine.

88
A great realm will be left desolated,
Near the Ebro an assembly will be formed:
The Pyrenees mountains will console him,
When in May lands will be trembling.

89
Feet and hands bound between two boats,
Face anointed with honey, and sustained with milk:
Wasps and flies, paternal love vexed,
Cup-bearer to falsify, Chalice tried.

90
The stinking abominable disgrace,
After the deed he will be congratulated:
The great excuse for not being favorable,
That Neptune will not be persuaded to peace.

91
Of the leader of the naval war,
Red one unbridled, severe, horrible whim,
Captive escaped from the elder one in the bale,
When there will be born a son to the great Agrippa.

92
Prince of beauty so comely,
Around his head a plot, the second deed betrayed:
The city to the sword in dust the face burnt,
Through too great murder the head of the King hated.

93
The greedy prelate deceived by ambition,
He will come to reckon nothing too much for him:
He and his messengers completely trapped,
He who cut the wood sees all in reverse.

94
A King will be angry with the see-breakers,
When arms of war will be prohibited:
The poison tainted in the sugar for the strawberries,
Murdered by waters, dead, saying land, land.

95
Calumny against the cadet by the detractor,
When enormous and warlike deeds will take place:
The least part doubtful for the elder one,
And soon in the realm there will be partisan deeds.

96
Great city abandoned to the soldiers,
Never was mortal tumult so close to it:
Oh, what a hideous calamity draws near,
Except one offense nothing will be spared it.

97
At forty-five degrees the sky will burn,
Fire to approach the great new city:
In an instant a great scattered flame will leap up,
When one will want to demand proof of the Normans.

98
Ruin for the Volcae so very terrible with fear,
Their great city stained, pestilential deed:
To plunder Sun and Moon and to violate their temples:
And to redden the two rivers flowing with blood.

99
The learned enemy will find himself confused,
His great army sick, and defeated by ambushes,
The Pyrenees and Pennine Alps will be denied him,
Discovering near the river ancient jugs.

100
INCANTATION OF THE LAW AGAINST INEPT CRITICS
Let those who read this verse consider it profoundly,

Let the profane and the ignorant herd keep away:
And far away all Astrologers, Idiots and Barbarians,
May he who does otherwise be subject to the sacred rite.

Epistle to Henry II

EPISTLE TO HENRY II

TO THE MOST INVINCIBLE
MOST POWERFUL AND MOST CHRISTIAN
HENRY, KING OF FRANCE THE SECOND:
MICHEL NOSTRADAMUS,
HIS VERY HUMBLE AND VERY OBEDIENT SERVANT AND SUBJECT,
WISHES VICTORY AND HAPPINESS

Ever since my long-beclouded face first presented itself before the immeasurable deity of your Majesty, O Most Christian and Most Victorious King, I have remained perpetually dazzled by that sovereign sight. I have never ceased to honor and venerate properly that date when I presented myself before a Majesty so singular and so humane. I have searched for some occasion on which to manifest high heart and stout courage, and thereby obtain even greater recognition of Your Most Serene Majesty. But I saw how obviously impossible it was for me to declare myself.

While I was seized with this singular desire to be transported suddenly from my long-beclouded obscurity to the illuminating presence of the first monarch of the universe, I was also long in doubt as to whom I would dedicate these last three Centuries of my prophecies, making up the thousand. After having meditated for a long time on an act of such rash audacity, I have ventured to address Your Majesty. I have not been daunted like those mentioned by that most grave author Plutarch, in his Life of Lycurgus, who were so astounded at the expense of the offerings and gifts brought as sacrifices to the temples of the immortal gods of that age, that they did not dare to present anything at all. Seeing your royal splendor to be accompanied by such an incomparable humanity, I have paid my address to it and not as those Kings of Persia whom one could neither stand before nor approach.

It is to a most prudent and most wise Prince that I have dedicated my nocturnal and prophetic calculations, which are composed rather out of a natural instinct, accompanied by a poetic furor, than according to the strict rules of poetry. Most of them have been integrated with astronomical calculations corresponding to the years, months and weeks of the regions, countries and most of the towns and cities of all Europe, including Africa and part of Asia, where most of all these coming events are to transpire. They are composed in a natural manner.

Indeed, someone, who would do well to blow his nose, may reply that the rhythm is as easy as the sense is difficult. That, O Most Humane king, is because most of the prophetic quatrains are so ticklish that there is no making way through them, nor is there any interpreting of them.

Nevertheless, I wanted to leave a record in writing of the years, towns, cities and regions in which most of the events will come to pass, even those of the year 1585 and of the year 1606, reckoning from the present time, which is March 14, 1557, and going far beyond to the events which will take place at the beginning of the seventh millenary, when, so far as my profound astronomical calculations and other knowledge have been able to make out, the adversaries of Jesus Christ and his Church will begin to multiply greatly.

I have calculated and composed all during choice hours of well-disposed days, and as accurately as I could, all when Minerva was free and not unfavorable. I have made computations for events over almost as long a period to come as that which has already passed, and by these they will know in all regions what is to happen in the course of time, just as it is written, with nothing superfluous added, although some may say, There can be no truth entirely determined concerning the future.

It is quite true, Sire, that my natural instinct has been inherited from my forebears, who did not believe in predicting, and that this is natural instinct has been adjusted and integrated with long calculations. At the same time, I freed my soul, mind and heart of all care, solicitude and vexation. All of these prerequisites for presaging I achieved in part by means of the brazen tripod.

There are some who would attribute to me that which is not mine at all. The eternal God alone, who is the thorough searcher of humane hearts, pious, just and merciful, is the true judge, and it is to him I pray to defend me from the calumny of evil men. These evil ones, in their slanderous way, would likewise want to inquire how all your most ancient progenitors, the Kings of France, have cured the scrofula, how those of other nations have cured the bite of snakes, how those of yet other nations have had a certain instinct for the art of divination and still others which would be too long to recite here.

Notwithstanding those who cannot contain the malignity of the evil spirit, as time elapses after my death, my writings will have more weight than during my lifetime. Should I, however, have made any errors in my calculation of dates, or prove unable to please everybody, I beg that your more than Imperial Majesty will forgive me. I protest before God and his Saints that I do not propose to insert any writings in this present Epistle that will be contrary to the true Catholic faith, whilst consulting the astronomical calculations to the best of my ability.

Such is the extent of time past, subject to correction by the most learned judgment, that the first man, Adam, came 1,242 years before Noah (not reckoning by such Gentile calculations as Varro used, but simply by the Holy Scriptures, as best my weak understanding and astronomical calculations can interpret them.) About 1,080 years after Noah and the universal flood came Abraham, who, according to some, was a first-rate astrologer and invented the Chaldean alphabet. About 515 or 516 years later came Moses, and from his time to that of David about 570 years elapsed. From the time of David to that of out Savior and Redeemer, Jesus Christ, born of the unique Virgin, 1,350 year elapsed, according to some chronographs. Some may object that this calculation cannot be true, because it differs from that of Eusebius. From the time of the human redemption to the detestable heresy of the Saracens about 621 years elapsed. From this one can easily add up the amount of time gone by.

Although my calculations may not hold good for all nations, they have, however, been determined by the celestial movements, combined with the emotion, handed down to me by my forebears, which comes over me at certain hours. But the danger of the times, O Most Serene King, requires that such secrets should not be bared except in enigmatic sentences having, however, only one sense and meaning, and nothing ambiguous or amphibological inserted. Rather they are under a cloudy obscurity, with a natural infusion not unlike the creation of the world, according to the calculation and Punic Chronicle of Joel: I will pour out my spirit upon all flesh and your sons and daughters will prophesy. But such Prophecy proceeded from the mouth of the Holy Ghost who was the sovereign and eternal power, together with the heavens, and caused some of them to predict great and marvelous events.

As for myself, I would never claim such a title, never, please God. I readily admit that all proceeds from God and render to Him thanks, honor and immortal praise. I have mixed therewith no divination coming from fate. All from God and nature, and for the most part integrated with celestial movements. It is much like seeing in a burning mirror, with clouded vision, the great events, sad, prodigious and calamitous events that in due time will fall upon the principal worshippers. First, upon the temples of God; secondly, upon those who, sustained by the earth, approach such a decadence. Also a thousand other calamitous events which will be known to happen in due time.

Writings of Nostradamus

For God will take notice of the long barrenness of the great dame, who thereupon will conceive two principal children. But she will be in danger, and the female to whom she will have given birth will also, because of the temerity of the age, be in danger of death in her eighteenth year, and will be unable to live beyond her thirty-sixth year. She will leave three males, and one female, and of these two will not have had the same father.

There will be great differences between the three brothers, and then there will be such great cooperation and agreement between them that the three and four parts of Europe will tremble. The youngest of them will sustain and augment the Christian monarchy, and under him sects will be elevated, and suddenly cast down, Arabs will be driven back, kingdoms united and new laws promulgated.

The oldest one will rule the land whose escutcheon is that of the furious crowned lions with their paws resting upon intrepid arms.

The one second in age, accompanied by the Latins, will penetrate far, until a second furious and trembling path has been beaten to the Great St. Bernard Pass. From there he will descend to mount the Pyrenees, which will not, however, be transferred to the French crown. And this third one will cause a great inundation of human blood, and for a long time Lent will not include March.

The daughter will be given for the preservation of the Christian Church. Her lord will fall into the pagan sect of the new infidels. Of her two children, one will be faithful to the Catholic Church, the other an infidel.

The unfaithful son, who, to his great confusion and later repentance, will want to ruin her, will have three widely scattered regions, namely, the Roman, Germany and Spain, which will set up diverse sects by armed force. The 50th to the 52th degree of latitude will be left behind.

And all will render the homage of ancient religions to the region of Europe north of the 48th parallel. The latter will have trembled first in vain timidity but afterwards the regions to its west, south and east will tremble. But the nature of their power will be such that what has been brought about by concord and union will prove insuperable by warlike conquests.

In nature they will be equal, but very different in faith.

After this the barren Dame, of greater power than the second, will be received by two of the nations. First, by them made obstinate by the onetime masters of the universe. Second, by the latter themselves.

The third people will extend their forces towards the circuit of the East of Europe where, in the Pannonias, they will be overwhelmed and slaughtered. By sea they will extend their Myrmidons and Germans to Adriatic Sicily. But they will succumb wholly and the Barbarian sect will be greatly afflicted and driven out by all the Latins.

Then the great Empire of the Antichrist will begin where once was Attila's empire and the new Xerxes will descend with great and countless numbers, so that the coming of the Holy Ghost, proceeding from the 48th degree, will make a transmigration, chasing out the abomination of the Christian Church, and whose reign will be for a time and to the end of time.

This will be preceded by a solar eclipse more dark and gloomy than any since the creation of the world, except that after the death and passion of Jesus Christ. And it will be in the month of October than the great translation will be made and it will be such that one will think the gravity of the earth has lost its natural movement and that it is to be plunged into the abyss of perpetual darkness.

Writings of Nostradamus

In the spring there will be omens, and thereafter extreme changes, reversals of realms and mighty earthquakes. These will be accompanied by the procreation of the new Babylon, miserable daughter enlarged by the abomination of the first holocaust. It will last for only seventy-three years and seven months.

Then there will issue from the stock which had remained barren for so long, proceeding from the 50th degree, one who will renew the whole Christian Church. A great place will be established, with union and concord between some of the children of opposite ideas, who have been separated by diverse realms. And such will be the peace that the instigator and promoter of military factions, born of the diversity of religions, will remain chained to the deepest pit. And the kingdom of the Furious One, who counterfeits the sage, will be united.

The countries, towns, cities, realms and provinces which will have abandoned their old customs to gain liberty, but which will in fact have enthralled themselves even more, will secretly have wearied of their liberty. Faith lost in their perfect religion, they will begin to strike to the left, only to return to the right. Holiness, for a long time overcome, will be replaced in accordance with the earliest writings.

Thereafter the great dog, the biggest of curs, will go forth and destroy all, the same old crimes being perpetrated again. Temples will be set up again as in ancient times, and the priest will be restored to his original position and he will begin his whoring and luxury, and will commit a thousand crimes.

At the eve of another desolation, when she is atop her most high and sublime dignity, some potentates and warlords will confront her, and take away her two swords, and leave her only the insignia, whose curvature attracts them. The people will make him go to the right and will not wish to submit themselves to those of the opposite extreme with the hand in acute position, who touch the ground, and want to drive spurs into them.

The people of the world from this benevolent slavery to which they had voluntary submitted. He will put himself under the protection of Mars, stripping Jupiter of all his honors and dignities, and establish himself in the free city in another scant Mesopotamia. The chief and governor will be cast out from the middle and hung up, ignorant of the conspiracy of one of the conspirators with the second Thrasibulus, who for a long time will have directed all this.

Then the impurities and abominations, with a great shame, will be brought out and manifested in the shadows of the veiled light, and will cease towards the end of the change in reign. The chiefs of the Church will be backward in the love of God, and several of them will apostatize from the true faith. Of the three sects, that which is in the middle, because of its own partisans, will be thrown a bit into decadence. The first one will be exterminated throughout all Europe and most of Africa by the third one, making use of the poor in spirit who, led by madmen to libidinous luxury, will adulterate.

The supporting common people will rise up and chase out the adherents of the legislators. From the way realms will have been weakened by the Easterners, it will seem that God the Creator has loosed Satan from the prisons of hell to give birth to the great Dog and Dogam, who will make such an abominable breach in the Churches that neither the reds nor the whites without eyes or hands will know what to make of it, and their power will be taken from them.

Then will commence a persecution of the Churches the like of which was never seen. Meanwhile, such a plague will arise that more than two thirds of the world will be removed. One will be unable to ascertain the true owners of fields and houses, and weeds growing in the streets of cities will rise higher than the knees. For the clergy there will be but utter desolation. The warlords will usurp what is returned from the City of the Sun, from Malta and the Isles of Hyhres. The great chain of the port which wakes its name from the marine ox will be opened.

Writings of Nostradamus

And a new incursion will be made by the maritime shores, wishing to deliver the Sierra Morea from the first Mahometan recapture. Their assaults will not all be in vain, and the place which was once the abode of Abraham will be assaulted by persons who hold the Jovialists in veneration. And this city of "Achem" will be surrounded and assailed on all sides by a most powerful force of warriors. Their maritime forces will be weakened by the Westerners, and great desolation will fall upon this realm. Its greatest cities will be depopulated and those who enter will fall under the vengeance of the wrath of God.

The sepulcher, for long an object of such great veneration, will remain in the open, exposed to the sight of the heavens, the Sun and the Moon. The holy place will be converted into a stable for a herd large and small, and used for profane purposes. Oh, what a calamitous affliction will pregnant women bear at this time.

For hereupon the principal Eastern chief will be vanquished by the Northerners and Westerners, and most of his people, stirred up, will be put to death, overwhelmed or scattered. His children, offspring of many women, will be imprisoned. Then will be accomplished the prophecy of the Royal Prophet, Let him hear the groaning of the captives, that he might deliver the children of those doomed to die.

What great oppression will then fall upon the Princes and Governors of Kingdoms, especially those which will be maritime and Eastern, whose tongues will be intermingled with all others: the tongue of the Latins, and of the Arabs, via the Phoenicians. And all these Eastern Kings will be chased, overthrown and exterminated, but not altogether, by means of the forces of the Kings of the North, and because of the drawing near of our age through the three secretly united in the search for death, treacherously laying traps for one another. This renewed Triumvirate will last for seven years, and the renown of this sect will extend around the world. The sacrifice of the hole and immaculate Wafer will be sustained.

Then the Lords of "Aquilon" [the North], two in number, will be victorious over the Easterners, and so great a noise and bellicose tumult will they make amongst them that all the East will tremble in terror of these brothers, yet not brothers, of "Aquilon" [the North].

By this discourse, Sire, I present these predictions almost with confusion, especially as to when they will take place. Furthermore, the chronology of time which follows conforms very little, if at all, with that which has already been set forth. Yet it was determined by astronomy and other sources, including Holy Scriptures, and thus could not err. If I had wanted to date each quatrain, I could have done so. But this would not have been agreeable to all, least of all to those interpreting them, and was not to be done until Your Majesty granted me full power to do so, lest calumniators be furnished with an opportunity to injure me.

Anyhow, I count the years from the creation of the world to the birth of Noah as 1,506, and from the birth of Noah to the completion of the Ark, at the time of the universal deluge, as 600 (let the years be solar, or lunar, or a mixture of the ten) I hold that the Sacred Scriptures use solar years. And at the end of these 600 years, Noah entered the Ark to be saved from the deluge. This deluge was universal, and lasted one year and two months. And 295 years elapsed from the end of the flood to the birth of Abraham, and 100 from then till the birth of Isaac. And 60 years later Jacob was born. 130 years elapsed between the time he entered Egypt and the time he came out. Between the entry of Jacob into Egypt and the exodus, 430 years passed. From the exodus to the building of the Temple by Solomon in the fourth year of his reign, 480 years. According to the calculations of the Sacred Writings, it was 490 years from the building of the Temple to the time of Jesus Christ. Thus, this calculation of mine, collected from the holy writ, comes to about 4,173 years and 8 months, more or less. Because there is such a diversity of sects, I will not go beyond Jesus Christ.

I have calculated the present prophecies according to the order of the chain which contains its revolution, all by astronomical doctrine modified by my natural instinct. After a while, I found the time when Saturn turns to enter on April 7 till August 25, Jupiter on June 14 till October 7, Mars from April 17 to June 22, Venus from April 9 to May 22, Mercury from February 3 to February 24. After that, from June 1 to June 24, and

Writings of Nostradamus

from September 25 to October 16, Saturn in Capricorn, Jupiter in Aquarius, Mars in Scorpio, Venus in Pisces, Mercury for a month in Capricorn, Aquarius and Pisces, the Moon in Aquarius, the Dragon's head in Libra: its tail in opposition following a conjunction of Jupiter and Mercury with a quadrature of Mars and Mercury, and the Dragon's head coinciding with a conjunction of the Sun and Jupiter. And the year without an eclipse peaceful.

But not everywhere. It will mark the commencement of what will long endure. For beginning with this year the Christian Church will be persecuted more fiercely than it ever was in Africa, and this will last up to the year 1792, which they will believe to mark a renewal of time.

After this the Roman people will begin to re-establish themselves, chasing away some obscure shadows and recovering a bit of their ancient glory. But this will not be without great division and continual changes. Thereafter Venice will raise its wings very high in great force and power, not far short of the might of ancient Rome.

At that time the great sails of Byzantium, allied with the Ligurians and through the support and power of "Aquilon" [the Northern Realm], will impede them so greatly that the two Cretans will be unable to maintain their faith. The arks built by the Warriors of ancient times will accompany them to the waves of Neptune. In the Adriatic great discord will arise, and that which will have been united will be separated. To a house will be reduced that which was, and is, a great city, including "Pampotamia" and "Mesopotamia" of Europe at 45, and others of 41, 42 and 37 degrees.

It will be at this time and in these countries that the infernal power will set the power of its adversaries against the Church of Jesus Christ. This will constitute of the second Antichrist, who will persecute that Church and its true Vicar, by means of the power of three temporal kings who in their ignorance will be seduced by tongues which, in the hands of the madmen, will cut more than any sword.

The said reign of the Antichrist will last only to the death of him who was born at the beginning of the age and of the other one of Lyon, associated with the elected one of the House of Modena and of Ferrara, maintained by the Adriatic Ligurians and the proximity of great Sicily. Then the Great St. Bernard will be passed.

The Gallic Ogmios will be accompanied by so great a number that the Empire of his great law will extend very far. For some time thereafter the blood of the Innocent will be shed profusely by the recently elevated guilty ones. Then, because of great floods, the memory of things contained in these instruments will suffer incalculable loss, even letters. This will happen to the "Aquiloners" [the Northern People] by the will of God.

Once again Satan will be bound, universal peace will be established among men, and the Church of Jesus Christ will be delivered from all tribulations, although the Philistines would like to mix in the honey of malice and their pestilent seduction. This will be near the seventh millenary, when the sanctuary of Jesus Christ will no longer be trodden down by the infidels who come from "Aquilon" [the North]. The world will be approaching a great conflagration, although, according to my calculations in my prophecies, the course of time runs much further.

In the Epistle that some years ago I dedicated to my son, Cisar Nostradamus, I declared some points openly enough, without presage. But here, Sire, are included several great and marvelous events which those to come after will see.

During this astrological supputation, harmonized with the Holy Scriptures, the persecution of the Ecclesiastical folk will have its origin in the power of the Kings of "Aquilon" [the North], united with the Easterners. This persecution will last for eleven years, or somewhat less, for then the chief King of "Aquilon" will fall.

Thereupon the same thing will occur in the South, where for the space of three years the Church people will be persecuted even more fiercely through the Apostatic seduction of one who will hold all the absolute power in the Church militant. The hole people of God, the observer of his law, will be persecuted fiercely and such will be their affliction that the blood of the true Ecclesiastics will flow everywhere.

One of the horrible temporal Kings will be told by his adherents, as the ultimate in praise, that he has shed more of human blood of Innocent Ecclesiastics than anyone else could have spilled of wine. This King will commit incredible crimes against the Church. Human blood will flow in the public streets and temples, like water after an impetuous rain, coloring the nearby rivers red with blood. The ocean itself will be reddened by another naval battle, such that one king will say to another, Naval battles have caused the sea to blush.

Then, in this same year, and in those following, there will ensue the most horrible pestilence, made more stupendous by the famine which will have preceded it. Such great tribulations will never have occurred since the first foundation of the Christian Church. It will cover all Latin regions, and will leave traces in some countries of the Spanish.

Thereupon the third King of "Aquilon" [the North], hearing the lament of the people of his principal title, will raise a very mighty army and, defying the tradition of his predecessors, will put almost everything back in its proper place, and the great Vicar of the hood will be put back in his former state. But desolated, and then abandoned by all, he will turn to find the Holy of Holies destroyed by paganism, and the old and new Testaments thrown out and burned.

After that Antichrist will be the infernal prince again, for the last time. All the Kingdoms of Christianity will tremble, even those of the infidels, for the space of twenty-five years. Wars and battles will be more grievous and towns, cities, castles and all other edifices will be burned, desolated and destroyed, with great effusion of vestal blood, violations of married woman and widows, and sucking children dashed and broken against the walls of towns. By means of Satan, Prince Infernal, so may evils will be committed that nearly all the world will find itself undone and desolated. Before these events, some rare birds will cry in the air: Hui, Hui [Today, today] and some time later will vanish.

After this has endured for a long time, there will be almost renewed another reign of Saturn, and golden age. Hearing the affliction of his people, God the Creator will command that Satan be cast into the depths of the bottomless pit, and bound there. Then a universal peace will commence between God and man, and Satan will remain bound for around a thousand years, and then all unbound.

All these figures represent the just integration of Holy Scriptures with visible celestial bodies, namely, Saturn, Jupiter, Mars and others conjoined, as can be seen at more length in some of the quatrains. I would have calculated more profoundly and integrated them even further, Most Serene King, but for the fact that some given to censure would raise difficulties. Therefore I withdraw my pen and seek nocturnal repose.

Many events, most powerful of all Kings, of the most astounding sort are to transpire soon, but I neither could nor would fit them all into this epistle; but in order to comprehend certain horrible facts, a few must be set forth. So great is your grandeur and humanity before men, and your piety before the gods, that you alone seem worthy of the great title of the Most Christian King, and to whom the highest authority in all religion should be deferred.

But I shall only beseech you, Most Clement King, by this singular and prudent humanity of yours, to understand rather the desire of my heart, and the sovereign wish I have to obey Your Most Serene Majesty, ever since my eyes approached your solar splendor, than the grandeur of my labor can attain to or acquire.

From Salon, this 27th of June, 1558.

Done by Michel Nostradamus at Salon-de-Crau in Provence.

Century VII

1
The arc of the treasure deceived by Achilles,
the quadrangle known to the procreators.
The invention will be known by the Royal deed;
a corpse seen hanging in the sight of the populace.

2
Opened by Mars Arles will not give war,
the soldiers will be astonished by night.
Black and white concealing indigo on land
under the false shadow you will see traitors sounded.

3
After the naval victory of France,
the people of Barcelona the Saillinons and those of Marseilles;
the robber of gold, the anvil enclosed in the ball,
the people of Ptolon will be party to the fraud.

4
The Duke of Langres besieged at Dile
accompanied by people from Autun and Lyons.
Geneva, Augsburg allied to those of Mirandola,
to cross the mountains against the people of Ancona.

5
Some of the wine on the table will be spilt,
the third will not have that which he claimed.
Twice descended from the black one of Parma,
Perouse will do to Pisa that which he believed.

6
Naples, Palerma and all of Sicily
will be uninhabited through Barbarian hands.
Corsica, Salerno and the island of Sardinia,
hunger, plague, war the end of extended evils.

7
Upon the struggle of the great light horses,
it will be claimed that the great crescent is destroyed.
To kill by night, in the mountains,
dressed in shepherdâs' clothing, red gulfs in the deep ditch.

8
Florense, flee, flee the nearest Roman,

at Fiesole will be conflict given:
blood shed, the greatest one take by the hand,
neither temple nor sex will be pardoned.

9
The lady in the absence of her great master
will be begged for love by the Viceroy.
Feigned promise and misfortune in love,
in the hands of the great Prince of Bar.

10
By the great Prince bordering Le Mans,
brave and valiant leader of the great army;
by land and sea with Bretons and Normans,
to pass Gibraltar and Barcelona to pillage the island.

11
eye, feet wounded rude disobedient;
strange and very bitter news to the lady;
more than five hundred of here people will be killed.

12
The great younger son will make an end of the war,
he assembles the pardoned before the gods;
Cahors and Moissac will go far from the prison,
a refusal at Lectoure, the people of Agen shaved.

13
From the marine tributary city,
the shaven head will take up the satrapy;
to chase the sordid man who will the be against him.
For fourteen years he will hold the tyranny.

14
He will come to expose the false topography,
the urns of the tombs will be opened.
Sect and holy philosophy to thrive,
black for white and the new for the old.

15
Before the city of the Insubrian lands,
for seven years the siege will be laid;
a very great king enters it,
the city is then free, away from its enemies.

16
The deep entry made by the great Queen
will make the place powerful and inaccessible;
the army of the three lions will be defeated
causing within a thing hideous and terrible.

17
The prince who has little pity of mercy
will come through death to change (and become) very knowledgeable.
The kingdom will be attended with great tranquillity,
when the great one will soon be fleeced.

18
The besieged will color their pacts,
but seven days later they will make a cruel exit:
thrown back inside, fire and blood, seven put to the ax
the lady who had woven the peace is a captive.

19
The fort at Nice will not engage in combat,
it will be overcome by shining metal.
This deed will be debated for a long time,
strange and fearful for the citizens.

20
Ambassadors of the Tuscan language
will cross the Alps and the sea in April and May.
The man of the calf will deliver an oration,
not coming to wipe out the French way of life.

21
By the pestilential enmity of Languedoc,
the tyrant dissimulated will be driven out.
The bargain will be made on the bridge at Sorgues
to put to death both him and his follower

22
The citizens of Mesopotamia
angry with their friends from Tarraconne;
games, rites, banquets, every person asleep,
the vicar at Rhine, the city taken and those of Ausonia.

23
The Royal scepter will be forced to take
that which his predecessors had pledged.
Because they do not understand about the ring
when they come to sack the palace.

24
He who was buried will come out of the tomb,
He will cause the fort of the bridge to be tied in chains:
Poisoned with the spawn of a pimp,
the great one from Lorraine by the Marquis du Pont.

25
Through long war all the army exhausted,
so that they do not find money for the soldiers;

instead of gold or silver, they will come to coin leather,
Gallic brass, and the crescent sign of the Moon.

26
Foists and galleys around seven ships,
a mortal war will be let loose.
The leader from Madrid will receive a wound from arrows,
two escaped and five brought to land.

27
At the wall of Vasto the great cavalry
are impeded by the baggage near Ferrara.
At Turin they will speedily commit such robbery
that in the fort they will ravish their hostage.

28
The captain will lead a great herd
on the mountain closest to the enemy.
Surrounded by fire he makes such a way,
all escape except for thirty put on the spit.

29
The great one of Alba will come to rebel,
he will betray his great forebears.
The great man of Guise will come to vanquish him,
led captive with a monument erected.

30
The sack approaches, fire and great bloodshed.
Po the great rivers, the enterprise for the clowns;
after a long wait from Genoa and Nice,
Fossano, Turin the capture at Savigliano.

31
From Languedoc and Guienne more than ten
thousand will want to cross the Alps again.
The great Savoyards march against Brindisi,
Aquino and Bresse will come to drive them back.

32
From the bank of Montereale will be born one
who bores and calculates becoming a tyrant.
To raise a force in the marches of Milan,
to drain Faenza and Florence of gold and men

33
The kingdom stripped of its forces by fraud,
the fleet blockaded, passages for the spy;
two false friends will come to rally
to awaken hatred for a long time dormant.

34
The French nation will be in great grief,
vain and lighthearted, they will believe rash things.
No bread, salt, wine nor water, venom nor ale,
the greater one captured, hunger, cold and want.

35
The great fish will come to complain and weep
for having chosen, deceived concerning his age:
he will hardly want to remain with them,
he will be deceived by those (speaking) his own tongue.

36
God, the heavens, all the divine words in the waves,
carried by seven red-shaven heads to Byzantium:
against the anointed three hundred from Trebizond,
will make two laws, first horror then trust.

37
Ten sent to put the captain of the ship to death,
are altered by one that there is open revolt in the fleet.
Confusion, the leader and another stab and bite each other
at Lerins and the Hyer s, ships, prow into the darkness.

38
The elder royal one on a frisky horse
will spur so fiercely that it will bolt.
Mouth, mouthful, foot complaining in the embrace;
dragged, pulled, to die horribly.

39
The leader of the French army
will expect to lose the main phalanx.
Upon the pavement of oats and slate
the foreign nation will be undermined through Genoa.

40
Within casks anointed outside with oil and grease
twenty-one will be shut before the harbor,
at second watch; through death they will do great deeds;
to win the gates and be killed by the watch.

41
The bones of the feet and the hands locked up,
because of the noise the house is uninhabited for a long time.
Digging in dreams they will be unearthed,
the house healthy in inhabited without noise.

42
Two newly arrived have seized the poison,
to pour it in the kitchen of the great Prince.

By the scullion both are caught in the act,
taken he who thought to trouble the elder with death.

Century VIII

1
Pau, Nay, Loron will be more of fire than blood,
to swim in praise, the great one to flee to the confluence (of rivers).
He will refuse entry to the magpies
Pampon and the Durance will keep them confined.

2
Condom and Auch and around Mirande,
I see fire from the sky which encompasses them.
Sun and Mars conjoined in Leo, then at Marmande,
lightning, great hail, a wall falls into the Garonne.

3
Within the strong castle of Vigilance and Resviers
the younger born of Nancy will be shut up.
In Turin the first ones will be burned,
when Lyons will be transported with grief.

4
The cock will be received into Monaco,
the Cardinal of France will appear;
He will be deceived by the Roman legation;
weakness to the eagle, strength will be born to the cock.

5
There will appear a shining ornate temple,
the lamp and the candle at Borne and Breteuil.
For the canton of Lucerne turned aside,
when one will see the great cock in his shroud.

6
Lighting and brightness are seen at Lyons shining,
Malta is taken, suddenly it will be extinguished.
Sardon, Maurice will act deceitfully,
Geneva to London, feigning treason towards the cock.

7
Vercelli, Milan will give the news,
the wound will be given at Pavia.
To run in the Seine, water, blood and fire through Florence,
the unique one falling from high to low calling for help.

8
Near Focia enclosed in some tuns
Chivasso will plot for the eagle.

The elected one driven out, he and his people shut up,
rape with Turin, the bride led away.

9
While the eagle is united with the cock at Savonna,
the Eastern Sea and Hungary.
The army at Naples, Palermo, the marches of Ancona,
Rome and Venice a great outcry by the Barbarian.

10
A great stench will come from Lausanne,
but they will not know its origin,
they will put out all people from distant places,
fire seen in the sky, a foreign nation defeated.

11
A multitude of people will appear at Vicenza
without force, fire to burn the Basilica.
Near Lunage the great one of Valenza defeated:
at a time when Venice takes up the quarrel through custom.

12
He will appear near to Buffalora
the highly born and tall one entered into Milan.
The Abbe of Foix with those of Saint–Meur
will cause damage dressed up as serfs.

13
The crusader brother through impassioned love
will cause Bellerophon to die through Proteus;
the fleet for a thousand years, the maddened woman,
the potion drunk, both of them then die.

14
The great credit of gold and abundance of silver
will cause honor to be blinded by lust;
the offense of the adulterer will become known,
which will occur to his great dishonor.

15
Great exertions towards the North by a man–woman
to vex Europe and almost all the Universe.
The two eclipses will be put into such a rout
that they will reinforce life or death for the Hungarians.

16
At the place where HIERON has his ship built,
there will be such a great sudden flood,
that one will not have a place nor land to fall upon,
the waters mount to the Olympic Fesulan.

17
Those at ease will suddenly be cast down,
the world put into trouble by three brothers;
their enemies will seize the marine city,
hunger, fire, blood, plague, all evils doubled.

18
The cause of her death will be issued from Florence,
one time before drunk by young and old;
by the three lilies they will give her a great pause.
Save through her offspring as raw meat is dampened.

19
To support the great troubled Cappe;
the reds will march in order to clarify it;
a family will be almost overcome by death,
the red, red ones will knock down the red one.

20
The false message about the rigged election
to run through the city stopping the broken pact;
voices bought, chapel stained with blood,
the empire contracted to another one.

21
Three foists will enter the port of Agde
carrying the infection and pestilence, not the faith.
Passing the bridge they will carry off a million,
the bridge is broken by the resistance of a third.

22
Coursan, Narbonne through the salt to warn
Tuchan, the grace of Perpignan betrayed;
the red town will not wish to consent to it,
in a high flight, a copy flag and a life ended.

23
Letters are found in the queen's chests,
no signature and no name of the author.
The ruse will conceal the offers;
so that they do not know who the lover is.

24
The lieutenant at the door of the house,
will knock down the great man of Perpignan.
Thinking to save himself at Montpertuis,
the bastard of Lusignan will be deceived.

25
The heart of the lover, awakened by furtive love
will ravish the lady in the stream.

She will pretend bashfully to be half injured,
the father of each will deprive the body of its soul.

26
The bones of Cato found in Barcelona,
placed, discovered, the site found again and ruined.
The great one who holds, but does not hold,
wants Pamplona, drizzle at the abbey of Montserrat.

27
The auxiliary way, one arch upon the other,
Le Muy deserted except for the brave one and his genet.
The writing of the Phoenix Emperor,
seen by him which is (shown) to no other.

28
The copies of gold and silver inflated,
which after the theft were thrown into the lake,
at the discovery that all is exhausted and dissipated by the debt.
All scrips and bonds will be wiped out.

29
At the fourth pillar which they dedicate to Saturn
split by earthquake and by flood;
under Saturn's building an urn is found
gold carried off by Caepio and then restored.

30
In Toulouse, not far from Beluzer
making a deep pit a palace of spectacle,
the treasure found will come to vex everyone
in two places and near the Basacle.

31
The first great fruit of the prince of Perchiera,
then will come a cruel and wicked man.
In Venice he will lose his proud glory,
and is led into evil by then younger Selin.

32
French king, beware of your nephew
who will do so much that your only son
will be murdered while making his vows to Venus;
accompanied at night by three and six.

33
The great one who will be born of Verona and Vincenza
who carries a very unworthy surname;
he who at Venice will wish to take vengeance,
himself taken by a man of the watch and sign.

34
After the victory of the Lion over the Lion,
there will be great slaughter on the mountain of Jura;
floods and dark-colored people of the seventh (of a million),
Lyons, Ulm at the mausoleum death and the tomb.

35
At the entrance to Garonne and Baise
and the forest not far from Damazan,
discoveries of the frozen sea, then hail and north winds.
Frost in the Dardonnais through the mistake of the month.

36
It will be committed against the anointed brought
from Lons le Saulnier, Saint Aubin and Bell'oeuvre.
To pave with marble taken from distant towers,
not to resist Bletteram and his masterpiece.

37
The fortress near the Thames
will fall when the king is locked up inside.
He will be seen in his shirt near the bridge,
one facing death then barred inside the fortress.

38
The King of Blois will reign in Avignon,
once again the people covered in blood.
In the Rhine he will make swim
near the walls up to five, the last one near Nolle.

39
He who will have been for the Byzantine prince
will be taken away by the prince of Toulouse.
The faith of Foix through the leader of Tolentino
will fail him, not refusing the bride.

40
The blood of the Just for Taur and La Duarade
in order to avenge itself against the Saturnines.
They will immerse the band in the new lake,
then they will march against Alba.

41
a fox will be elected without speaking one word,
appearing saintly in public living on barley bread,
afterwards he will suddenly become a tyrant
putting his foot on the throats of the greatest men.

42
Through avarice, through force and violence
the chief of Orleans will come to vex his supporters.

Near St. Memire, assault and resistance.
Dead in his tent they will say he is asleep inside.

43
Through the fall of two bastard creatures
the nephew of the blood will occupy the throne.
Within Lectoure there will be blows of lances,
the nephew through fear will fold up his standard.

44
The natural offspring off Ogmios
will turn off the road from seven to nine.
To the king long friend of the half man,
Navarre must destroy the fort at Pau.

45
With his hand in a sling and his leg bandaged,
the younger brother of Calais will reach far.
At the word of the watch, the death will be delayed,
then he will bleed at Easter in the Temple.

46
Paul the celibate will die three leagues from Rome,
the two nearest flee the oppressed monster.
When Mars will take up his horrible throne,
the Cock and the Eagle, France and the three brothers.

47
Lake Trasimene will bear witness
of the conspirators locked up inside Perugia.
A fool will imitate the wise one,
killing the Teutons, destroying and cutting to pieces.

48.
Saturn in Cancer, Jupiter with Mars
in February Chaldondon'salva tierra.
Sierra Morena besieged on three sides
near Verbiesque, war and mortal conflict.

49
Saturn in Taurus, Jupiter in Aquarius. Mars in Sagittarius,
the sixth of February brings death.
Those of Tardaigne so great a breach at Bruges,
that the barbarian chief will die at Ponteroso.

50
The plague around Capellades,
another famine is near to Sagunto;
the knightly bastard of the good old man
will cause the great one of Tunis to lose his head.

51
The Byzantine makes an oblation
after having taken back Cordoba.
A long rest on his road, the vines cut down,
at sea the passing prey captured by the Pillar.

52 ---- Unfinished/Censored ----
The king of Blois to reign in Avignon,
from Amboise and Seme the length of the Indre:
claws at Poitiers holy wings ruined
before Boni. . . .

53
Within Boulogne he will want to wash away his misdeeds,
he cannot at the temple of the Sun.
He will fly away, doing very great things:
In the hierarchy he had never an equal.

54
Under the color of the marriage treaty,
a magnanimous act by the Chyren Selin:
St. Quintin and Arras recovered on the journey;
By the Spanish a second butcher's bench is made.

55
He will find himself shut in between two rivers,
casks and barrels joined to cross beyond:
eight bridges broken, their chief run through so many times,
perfect children's throats slit by the knife.

56
The weak band will occupy the land,
those of high places will make dreadful cries.
The large herd of the outer corner troubled,
near Edinburgh it falls discovered by the writings.

57
From simple soldier he will attain to Empire,
from the short robe he will grow into the long.
Brave in arms, much worse towards the Church,
he vexes the priests as water fills a sponge.

58
A kingdom divided by two quarreling brothers
to take the arms and the name of Britain.
The Anglican title will be advised to watch out,
surprised by night (the other is), led to the French air.

59
Twice put up and twice cast down,
the East will also weaken the West.

Its adversary after several battles
chased by sea will fail at time of need.

60
First in Gaul, first in Romania,
over land and sea against the English and Paris.
Marvelous deeds by that great troop,
violent, the wild beast will lose Lorraine.

61
Never by the revelation of daylight
will he attain the mark of the scepter bearer.
Until all his sieges are at rest,
bringing to the Cock the gift of the armed legion.

62
When one sees the holy temple plundered,
the greatest of the Rhine profaning their sacred things;
because of them a very great pestilence will appear,
the king, unjust, will not condemn them.

63
When the adulterer wounded without a blow
will have murdered his wife and son out of spite;
his wife knocked down, he will strangle the child;
eight captives taken, choked beyond help.

64
The infants transported into the islands,
two out of seven will be in despair.
Those of the soil will be supported by it,
the name 'shovel' taken, the hope of the leagues fails.

65
The old man disappointed in his main hope,
will attain to the leadership of his Empire.
Twenty months he will hold rule with great force,
a tyrant, cruel, giving way to one worse.

66
When the inscription D.M. is found
in the ancient cave, revealed by a lamp.
Law, the King and Prince Ulpian tried,
the Queen and Duke in the pavilion under cover.

67
Paris, Carcassone, France to ruin in great disharmony,
neither one nor the other will be elected.
France will have the love and good will of the people,
Ferara, Colonna great protection.

68
The old Cardinal is deceived by the young one,
he will find himself disarmed, out of his position:
Do not show, Arles, that the double is perceived,
both Liqueduct and the Prince embalmed.

69
Beside the young one the old angel falls,
and will come to rise above him at the end;
ten years equal to most the old one falls again,
of three two and one, the eighth seraphim.

70
He will enter, wicked, unpleasant, infamous,
tyrannizing over Mesopotamia.
All friends made by the adulterous lady,
the land dreadful and black of aspect.

71
The number of astrologers will grow so great,
that they will be driven out, banned and their books censored.
In the year 1607 by sacred assemblies
so that none will be safe from the holy ones.

72
Oh what a huge defeat on the Perugian battlefield
and the conflict very close to Ravenna.
A holy passage when they will celebrate the feast,
the conqueror banished to eat horse meat.

73
The king is struck by a barbarian soldier,
unjustly, not far from death.
The greedy will be the cause of the deed,
conspirator and realm in great remorse.

74
A king entered very far into the new land
while the subjects will come to bid him welcome;
his treachery will have such a result
that to the citizens it is a reception instead of a festival.

75
The father and son will be murdered together,
the leader within his pavilion.
The mother at Tours will have her belly swollen with a son,
a verdure chest with little pieces of paper.

76
More of a butcher than a king in England,
born of obscure rank will gain empire through force.

Coward without faith, without law he will bleed the land;
His time approaches so close that I sigh.

77
The antichrist very soon annihilates the three,
twenty-seven years his war will last.
The unbelievers are dead, captive, exiled;
with blood, human bodies, water and red hail covering the earth.

78
A soldier of fortune with twisted tongue
will come to the sanctuary of the gods.
He will open the door to heretics
and raise up the Church militant.

79
He who loses his father by the sword, born in a Nunnery,
upon this Gorgon's blood will conceive anew;
in a strange land he will do everything to be silent,
he who will burn both himself and his child.

80
The blood of innocents, widow and virgin,
so many evils committed by means of the Great Red One,
holy images placed over burning candles,
terrified by fear, none will be seen to move.

81
The new empire in desolation
will be changed from the Northern Pole.
From Sicily will come such trouble that
it will bother the enterprise tributary to Philip.

82
Thin tall and dry, playing the good valet
in the end will have nothing but his dismissal;
sharp poison and letters in his collar,
he will be seized escaping into danger.

83
The largest sail set out of the port of Zara,
near Byzantium will carry out its enterprise.
Loss of enemy and friend will not be,
a third will turn on both with great pillage and capture.

84
Paterno will hear the cry from Sicily,
all the preparations in the Gulf of Trieste;
it will be heard as far as Sicily
flee oh, flee, so may sails, the dreaded pestilence !

85
Between Bayonne and St. Jean de Luz
will be placed the promontory of Mars.
To the Hanix of the North, Nanar will remove the light,
then suffocate in bed without assistance.

86
Through Emani, Tolosa and Villefranche,
an infinite band through the mountains of Adrian.
Passes the river, Cambat over the plank for a bridge,
Bayonne will be entered all crying Bigoree.

87
A death conspired will come to its full effect,
the charge given and the voyage of death.
Elected, created, received (then) defeated by its followers,
in remorse the blood of innocence in front of him.

88
A noble king will come to Sardinia,
who will only rule for three years in the kingdom.
He will join with himself several colors;
he himself, after taunts, care spoils slumber.

89
In order not to fall into the hands of his uncle
who slaughtered his children in order to reign.
Pleasing with the people, putting his foot on Peloncle,
dead and dragged between armored horses.

90
When those of the cross are found their senses troubled,
in place of sacred things he will see a horned bull,
through the virgin the pig's place will then be filled,
order will no longer be maintained by the king.

91
Entered among the field of the Rhine
where those of the cross are almost united,
the two lands meeting in Pisces
and a great number punished by the flood.

92
Far distant from his kingdom, sent on a dangerous journey,
he will lead a great army and keep it for himself.
The king will hold his people captive and hostage,
he will plunder the whole country on his return.

93
For seven months, no longer, will he hold the office of prelate,
through his death a great schism will arise;

for seven months another acts as prelate near Venice,
peace and union are reborn.

94
In front of the lake where the dearest one was destroyed
for seven months and his army routed;
Spaniards will be devastating by means of Alba,
through delay in giving battle, loss.

95
The seducer will be placed in a ditch
and will be tied up for some time.
The scholar joins the chief with his cross.
The sharp right will draw the contented ones.

96
The sterile synagogue without any fruit,
will be received by the infidels,
the daughter of the persecuted (man) of Babylon,
miserable and sad, they will clip her wings.

97
At the end of the Var the great powers change;
near the bank three beautiful children are born.
Ruin to the people when they are of age;
in the country the kingdom is seen to grow and change more.

98
Of the church men the blood will be poured forth
as abundant as water in (amount);
for a long time it will not be restrained,
woe, woe, for the clergy ruin and grief.

99
Through the power of three temporal kings,
the sacred seat will be put in another place,
where the substance of the body and the spirit
will be restored and received as the true seat.

100
By the great number of tears shed,
from top to bottom and from the bottom to the very top,
a life is lost through a game with too much faith,
to die of thirst through a great deficiency.

Century IX

1
In the house of the translator of Bourg,
The letters will be found on the table,

One-eyed, red-haired, white, hoary-headed will hold the course,
Which will change for the new Constable.

2
From the top of the Aventine hill a voice heard,
Be gone, be gone all of you on both sides:
The anger will be appeased by the blood of the red ones,
From Rimini and Prato, the Colonna expelled.

3
The "great cow" at Racenna in great trouble,
Led by fifteen shut up at Fornase:
At Rome there will be born two double-headed monsters,
Blood, fire, flood, the greatest ones in space.

4
The following year discoveries through flood,
Two chiefs elected, the first one will not hold:
The refuge for the one of them fleeing a shadow,
The house of which will maintain the first one plundered.

5
The third toe will seem first
To a new monarch from low high,
He who will possess himself as a Tyrant of Pisa and Lucca,
To correct the fault of his predecessor.

6
An infinity of Englishmen in Guienne
Will settle under the name of Anglaquitaine:
In Languedoc, Ispalme, Bordelais,
Which they will name after Barboxitaine.

7
He who will open the tomb found,
And will come to close it promptly,
Evil will come to him, and one will be unable to prove,
If it would be better to be a Breton or Norman King.

8
The younger son made King will put his father to death,
After the conflict very dishonest death:
Inscription found, suspicion will bring remorse,
When the wolf driven out lies down ion the bedstead.

9
When the lamp burning with inextinguishable fire
Will be found in the temple of the Vestals:
Child found in fire, water passing through the sieve:
To perish in water N"mes, Toulouse the markets to fall.

10
The child of a monk and nun exposed to death,
To die through a she–bear, and carried off by a boar,
The army will be camped by Foix and Pamiers,
Against Toulouse Carcassonne the harbinger to form.

11
Wrongly will they come to put the just one to death,
In public and in the middle extinguished:
So great a pestilence will come to arise in this place,
That the judges will be forced to flee.

12
So much silver of Diana and Mercury,
The images will be found in the lake:
The sculptor looking for new clay,
He and his followers will be steeped in gold.

13
The exiles around Sologne,
Led by night to march into Auxois,
Two of Modena for Bologna cruel,
Placed discovered by the fire of Buzanais.

14
Dyers' caldrons put on the flat surface,
Wine, honey and oil, and built over furnaces:
They will be immersed, innocent, pronounced malefactors,
Seven of Bordeaux smoke still in the cannon.

15
Near Perpignan the red ones detained,
Those of the middle completely ruined led far off:
Three cut in pieces, and five badly supported,
For the Lord and Prelate of Burgundy.

16
Out of Castelfranco will come the assembly,
The ambassador not agreeable will cause a schism:
Those of Riviera will be in the squabble,
And they will refuse entry to the great gulf.

17
The third one first does worse than Nero,
How much human blood to flow, valiant, be gone:
He will cause the furnace to be rebuilt,
Golden Age dead, new King great scandal.

18
The lily of the Dauphin will reach into Nancy,
As far as Flanders the Elector of the Empire:

New confinement for the great Montmorency,
Outside proven places delivered to celebrated punishment.

19
In the middle of the forest of Mayenne,
Lightning will fall, the Sun in Leo:
The great bastard issued from the great one Maine,
On this day a point will enter the blood of Foug res.

20
By night will come through the forest of Reines,
Two couples roundabout route Queen the white stone,
The monk king in gray in Varennes:
Elected Capet causes tempest, fire, blood, slice.

21
At the tall temple of Saint–Solenne at Blois,
Night Loire bridge, Prelate, King killing outright:
Crushing victory in the marshes of the pond,
Whence prelacy of whites miscarrying.

22
The King and his court in the place of cunning tongue,
Within the temple facing the palace:
In the garden the Duke of Mantua and Alba,
Alba and Mantua dagger tongue and palace.

23
The younger son playing outdoors under the arbor,
The top of the roof in the middle on his head,
The father King in the temple of Saint–Solonne,
Sacrificing he will consecrate festival smoke.

24
Upon the palace at the balcony of the windows,
The two little royal ones will be carried off:
To pass Orleans, Paris, abbey of Saint–Denis,
Nun, wicked ones to swallow green pits.

25
Crossing the bridges to come near the Roisiers,
Sooner than he thought, he arrived late.
The new Spaniards will come to Beziers,
So that this chase will break the enterprise.

26
Departed by the bitter letters the surname of Nice,
The great Cappe will present something, not his own;
Near Voltai at the wall of the green columns,
After Piombino the wind in good earnest.

27
The forester, the wind will be close around the bridge,
Received highly, he will strike the Dauphin.
The old craftsman will pass through the woods in a company,
Going far beyond the right borders of the Duke.

28
The Allied fleet from the port of Marseilles,
In Venice harbor to march against Hungary.
To leave from the gulf and the bay of Illyria,
Devastation in Sicily, for the Ligurians, cannon shot.

29
When the man will give way to none,
Will wish to abandon a place taken, yet not taken;
Ship afire through the swamps, bitumen at Charlieu,
St. Quintin and Calais will be recaptured.

30
At the port of Pola and of San Nicolo,
A Normand will punish in the Gulf of Quarnero:
Capet to cry alas in the streets of Byzantium,
Help from Cadiz and the great Philip.

31
The tin island of St. George half sunk;
Drowsy with peace, war will arise,
At Easter in the temple abysses opened.

32
A deep column of fine porphyry is found,
Inscriptions of the Capitol under the base;
Bones, twisted hair, the Roman strength tried,
The fleet is stirred at the harbor of Mitylene.

33
Hercules King of Rome and of "Annemark,"
With the surname of the chief of triple Gaul,
Italy and the one of St. Mark to tremble,
First monarch renowned above all.

34
The single part afflicted will be mitered,
Return conflict to pass over the tile:
For five hundred one to betray will be titled
Narbonne and Salces we have oil for knives.

35
And fair Ferdinand will be detached,
To abandon the flower, to follow the Macedonian:
In the great pinch his course will fail,

And he will march against the Myrmidons.

36
A great King taken by the hands of a young man,
Not far from Easter confusion knife thrust:
Everlasting captive times what lightning on the top,
When three brothers will wound each other and murder.

37
Bridge and mills overturned in December,
The Garonne will rise to a very high place:
Walls, edifices, Toulouse overturned,
So that none will know his place like a matron.

38
The entry at Blaye for La Rochelle and the English,
The great Macedonian will pass beyond:
Not far from Agen will wait the Gaul,
Narbonne help beguiled through conversation.

39
In Albisola to Veront and Carcara,
Led by night to seize Savona:
The quick Gascon La Turbie and L'Escar ne:
Behind the wall old and new palace to seize.

40
Near Saint–Quintin in the forest deceived,
In the Abbey the Flemish will be cut up:
The two younger sons half–stunned by blows,
The rest crushed and the guard all cut to pieces.

41
The great "Chyren" will seize Avignon,
From Rome letters in honey full of bitterness:
Letter and embassy to leave from Chanignon,
Carpentras taken by a black duke with a red feather.

42
From Barcelona, from Genoa and Venice,
From Sicily pestilence Monaco joined:
They will take their aim against the Barbarian fleet,
Barbarian driven 'way back as far as Tunis.

43
On the point of landing the Crusader army
Will be ambushed by the Ishmaelites,
Struck from all sides by the ship Impetuosity,
Rapidly attacked by ten elite galleys.

44
Leave, leave Geneva every last one of you,
Saturn will be converted from gold to iron,
Raypoz will exterminate all who oppose him,
Before the coming the sky will show signs.

45
None will remain to ask,
Great Mendosus will obtain his dominion:
Far from the court he will cause to be countermanded
Piedmont, Picardy, Paris, Tuscany the worst.

46
Be gone, flee from Toulouse ye red ones,
For the sacrifice to make expiation:
The chief cause of the evil under the shade of pumpkins:
Dead to strangle carnal prognostication.

47
The undersigned to an infamous deliverance,
And having contrary advice from the multitude:
Monarch changes put in danger over thought,
Shut up in a cage they will see each other face to face.

48
The great city of the maritime Ocean,
Surrounded by a crystalline swamp:
In the winter solstice and the spring,
It will be tried by frightful wind.

49
Ghent and Brussels will march against Antwerp,
The Senate of London will put to death their King:
Salt and wine will overthrow him,
To have them the realm turned upside down.

50
Mendosus will soon come to his high realm,
Putting behind a little the Lorrainers:
The pale red one, the male in the interregnum,
The fearful youth and Barbaric terror.

51
Against the red ones sects will conspire,
Fire, water, steel, rope through peace will weaken:
On the point of dying those who will plot,
Except one who above all the world will ruin.

52
Peace is nigh on one side, and war,
Never was the pursuit of it so great:

To bemoan men, women innocent blood on the land,
And this will be throughout all France.

53
The young Nero in the three chimneys
Will cause live pages to be thrown to burn:
Happy those who will be far away from such practices,
Three of his blood will have him ambushed to death.

54
There will arrive at Porto Corsini,
Near Ravenna, he who will plunder the lady:
In the deep sea legate from Lisbon,
Hidden under a rock they will carry off seventy souls.

55
The horrible war which is being prepared in the West,
The following year will come the pestilence
So very horrible that young, old, nor beast,
Blood, fire Mercury, Mars, Jupiter in France.

56
The army near Houdan will pass Goussainville,
And at Maiotes it will leave its mark:
In an instant more than a thousand will be converted,
Looking for the two to put them back in chain and firewood.

57
In the place of Drux a King will rest,
And will look for a law changing Anathema:
While the sky will thunder so very loudly,
New entry the King will kill himself.

58
On the left side at the spot of Vitry,
The three red ones of France will be awaited:
All felled red, black one not murdered,
By the Bretons restored to safety.

59
At La Ferté-Vidame he will seize,
Nicholas held red who had produced his life:
The great Louise who will act secretly one will be born,
Giving Burgundy to the Bretons through envy.

60
Conflict Barbarian in the black Headdress,
Blood shed, Dalmatia to tremble:
Great Ishmael will set up his promontory,
Frogs to tremble Lusitania aid.

61
The plunder made upon the marine coast,
In Cittanova and relatives brought forward:
Several of Malta through the deed of Messina
Will be closely confined poorly rewarded.

62
To the great one of Ceramon-agora,
The crusaders will all be attached by rank,
The long-lasting Opium and Mandrake,
The Raugon will be released on the third of October.

63
Complaints and tears, cries and great howls,
Near Narbonne at Bayonne and in Foix:
Oh, what horrible calamities and changes,
Before Mars has made several revolutions.

64
The Macedonian to pass the Pyrenees mountains,
In March Narbonne will not offer resistance:
By land and sea he will carry on very great intrigue,
Capetian having no land safe for residence.

65
He will come to go into the corner of Luna,
Where he will be captured and put in a strange land:
The unripe fruits will be the subject of great scandal,
Great blame, to one great praise.

66
There will be peace, union and change,
Estates, offices, low high and high very low:
To prepare a trip, the first offspring torment,
War to cease, civil process, debates.

67
From the height of the mountains around the Is re,
One hundred assembled at the haven in the rock Valence:
From Ch%oteauneuf, Pierrelatte, in Donz re,
Against Crest, Romans, faith assembled.

68
The noble of Mount Aymar will be made obscure,
The evil will come at the junction of the Saine and Rhine:
Soldiers hidden in the woods on Lucy's day,
Never was there so horrible a throne.

69
One the mountain of Saint-Bel and L'Arbresle
The proud one of Grenoble will be hidden:

Beyond Lyons and Vienne on them a very great hail,
Lobster on the land not a third thereof will remain.

70
Sharp weapons hidden in the torches.
In Lyons, the day of the Sacrament,
Those of Vienne will all be cut to pieces,
By the Latin Cantons M‰con does not lie.

71
At the holy places animals seen with hair,
With him who will not dare the day:
At Carcassonne propitious for disgrace,
He will be set for a more ample stay.

72
Again will the holy temples be polluted,
And plundered by the Senate of Toulouse:
Saturn two three cycles completed,
In April, May, people of new leaven.

73
The Blue Turban King entered into Foix,
And he will reign less than an evolution of Saturn:
The White Turban King Byzantium heart banished,
Sun, Mars and Mercury near Aquarius.

74
In the city of Fertsod homicide,
Deed, and deed many oxen plowing no sacrifice:
Return again to the honors of Artemis,
And to Vulcan bodies dead ones to bury.

75
From Ambracia and the country of Thrace
People by sea, evil and help from the Gauls:
In Provence the perpetual trace,
With vestiges of their custom and laws.

76
With the rapacious and blood-thirsty king,
Issued from the pallet of the inhuman Nero:
Between two rivers military hand left,
He will be murdered by Young Baldy.

77
The realm taken the King will conspire,
The lady taken to death ones sworn by lot:
They will refuse life to the Queen and son,
And the mistress at the fort of the wife.

78
The Greek lady of ugly beauty,
Made happy by countless suitors:
Transferred out to the Spanish realm,
Taken captive to die a miserable death.

79
The chief of the fleet through deceit and trickery
Will make the timid ones come out of their galleys:
Come out, murdered, the chief renouncer of chrism,
Then through ambush they will pay him his wages.

80
The Duke will want to exterminate his followers,
He will send the strongest ones to strange places:
Through tyranny to ruin Pisa and Lucca,
Then the Barbarians will gather the grapes without vine.

81
The crafty King will understand his snares,
Enemies to assail from three sides:
A strange number tears from hoods,
The grandeur of the translator will come to fail.

82
By the flood and fierce pestilence,
The great city for long besieged:
The sentry and guard dead by hand,
Sudden capture but none wronged.

83
Sun twentieth of Taurus the earth will tremble very mightily,
It will ruin the great theater filled:
To darken and trouble air, sky and land,
Then the infidel will call upon God and saints.

84
The King exposed will complete the slaughter,
After having discovered his origin:
Torrent to open the tomb of marble and lead,
Of a great Roman with Medusine device.

85
To pass Guienne, Languedoc and the Rhine,
From Agen holding Marmande and La Reole:
To open through faith the wall, Marseilles will hold its throne,
Conflict near Saint–Paul–de–Mausole.

86
From Bourg–la–Reine they will come straight to Chartres,
And near Pont d'Antony they will pause:

Seven crafty as Martens for peace,
Paris closed by an army they will enter.

87
In the forest cleared of the Tuft,
By the hermitage will be placed the temple:
The Duke of ƒtampes through the ruse he invented
Will teach a lesson to the prelate of Montlhery.

88
Calais, Arras, help to Therouanne,
Peace and semblance the spy will simulate:
The soldiery of Savoy to descend by Roanne,
People who would end the rout deterred.

89
For seven years fortune will favor Philip,
He will beat down again the exertions of the Arabs:
Then at his noon perplexing contrary affair,
Young Ogmios will destroy his stronghold.

90
A captain of Great Germany
Will come to deliver through false help
To the King of Kings the support of Pannonia,
So that his revolt will cause a great flow of blood.

91
The horrible plague Perinthus and Nicopolis,
The Peninsula and Macedonia will it fall upon:
It will devastate Thessaly and Amphipolis,
An unknown evil, and from Anthony refusal.

92
The King will want to enter the new city,
Through its enemies they will come to subdue it:
Captive free falsely to speak and act,
King to be outside, he will keep far from the enemy.

93
The enemies very far from the fort,
The bastion brought by wagons:
Above the walls of Bourges crumbled,
When Hercules the Macedonian will strike.

94
Weak galleys will be joined together,
False enemies the strongest on the rampart:
Weak ones assailed Bratislava trembles,
LŸbeck and Meissen will take the barbarian side.

95
The newly made one will lead the army,
Almost cut off up to near the bank:
Help from the Milanais elite straining,
The Duke deprived of his eyes in Milan in an iron cage.

96
The army denied entry to the city,
The Duke will enter through persuasion:
The army led secretly to the weak gates,
They will put it to fire and sword, effusion of blood.

97
The forces of the sea divided into three parts,
The second one will run out of supplies,
In despair looking for the Elysian Fields,
The first ones to enter the breach will obtain the victory.

98
Those afflicted through the fault of a single one stained,
The transgressor in the opposite party:
He will send word to those of Lyons that compelled
They be to deliver the great chief of Molite.

99
The "Aquilon" Wind will cause the siege to be raised,
Over the walls to throw ashes, lime and dust:
Through rain afterwards, which will do them much worse,
Last help against their frontier.

100
Naval battle night will be overcome,
Fire in the ships to the West ruin:
New trick, the great ship colored,
Anger to the vanquished, and victory in a drizzle.

Century X

1
To the enemy, the enemy faith promised
Will not be kept, the captives retained:
One near death captured, and the remainder in their shirts,
The remainder damned for being supported.

2
The ship's veil will hide the sail galley,
The great fleet will come the lesser one to go out:
Ten ships near will turn to drive it back,
The great one conquered the united ones to join to itself.

3
After that five will not put out the flock,
A fugitive for Penelon he will turn loose:
To murmur falsely then help to come,
The chief will then abandon the siege.

4
At midnight the leader of the army
Will save himself, suddenly vanished:
Seven years later his reputation unblemished,
To his return they will never say yes.

5
Albi and Castres will form a new league,
Nine Arians Lisbon and the Portuguese:
Carcassonne and Toulouse will end their intrigue,
When the chief new monster from the Lauraguais.

6
The Gardon will flood N"mes so high
That they will believe Deucalion reborn:
Into the colossus the greater part will flee,
Vesta tomb fire to appear extinguished.

7
The great conflict that they are preparing for Nancy,
The Macedonian will say I subjugate all:
The British Isle in anxiety over wine and salt,
"Hem. mi." Philip two Metz will not hold for long.

8
With forefinger and thumb he will moisten the forehead,
The Count of Senigallia to his own son:
The Venus through several of thin forehead,
Three in seven days wounded dead.

9
In the Castle of Figueras on a misty day
A sovereign prince will be born of an infamous woman:
Surname of breeches on the ground will make him posthumous,
Never was there a King so very bad in his province.

10
Stained with murder and enormous adulteries,
Great enemy of the entire human race:
One who will be worse than his grandfathers, uncles or fathers,
In steel, fire, waters, bloody and inhuman.

11
At the dangerous passage below Junquera,
The posthumous one will have his band cross:

To pass the Pyrenees mountains without his baggage,
From Perpignan the duke will hasten to Tende.

12
Elected Pope, as elected he will be mocked,
Suddenly unexpectedly moved prompt and timid:
Through too much goodness and kindness provoked to die,
Fear extinguished guides the night of his death.

13
Beneath the food of ruminating animals,
led by them to the belly of the fodder city:
Soldiers hidden, their arms making a noise,
Tried not far from the city of Antibes.

14
Urnel Vaucile without a purpose on his own,
Bold, timid, through fear overcome and captured:
Accompanied by several pale whores,
Convinced in the Carthusian convent at Barcelona.

15
Father duke old in years and choked by thirst,
On his last day his don denying him the jug:
Into the well plunged alive he will come up dead,
Senate to the thread death long and light.

16
Happy in the realm of France, happy in life,
Ignorant of blood, death, fury and plunder:
For a flattering name he will be envied,
A concealed King, too much faith in the kitchen.

17
The convict Queen seeing her daughter pale,
Because of a sorrow locked up in her breast:
Lamentable cries will come then from Angoul me,
And the marriage of the first cousin impeded.

18
The house of Lorraine will make way for Vendime,
The high put low, and the low put high:
The son of Mammon will be elected in Rome,
And the two great ones will be put at a loss.

29
The day that she will be hailed as Queen,
The day after the benediction the prayer:
The reckoning is right and valid,
Once humble never was one so proud.

20
All the friend who will have belonged to the party,
For the rude in letters put to death and plundered:
Property up for sale at fixed price the great one annihilated.
Never were the Roman people so wronged.

21
Through the spite of the King supporting the lesser one,
He will be murdered presenting the jewels to him:
The father wishing to impress nobility on the son
Does as the Magi did of yore in Persia.

22
For not wishing to consent to the divorce,
Which then afterwards will be recognized as unworthy:
The King of the Isles will be driven out by force,
In his place put one who will have no mark of a king.

23
The remonstrances made to the ungrateful people,
Thereupon the army will seize Antibes:
The complaints will place Monace in the arch,
And at Frejus the one will take the shore from the other

24
The captive prince conquered in Italy
Will pass Genoa by sea as far as Marseilles:
Through great exertion by the foreigners overcome,
Safe from gunshot, barrel of bee's liquor.

25
Through the Ebro to open the passage of Bisanne,
Very far away will the Tagus make a demonstration:
In Pelligouxe will the outrage be committed,
By the great lady seated in the orchestra.

26
The successor will avenge his brother-in-law,
To occupy the realm under the shadow of vengeance:
Obstacle slain his blood for the death blame,
For a long time will Brittany hold with France.

27
Through the fifth one and a great Hercules
They will come to open the temple by hand of war:
One Clement, Julius and Ascanius set back,
The sword, key, eagle, never was there such a great animosity.

28
Second and third which make prime music
By the King to be sublimated in honor:

Through the fat and the thin almost emaciated,
By the false report of Venus to be debased.

29
In a cave of Saint–Paul–de–Mausole a goat
Hidden and seized pulled out by the beard:
Led captive like a mastiff beast
By the Bigorre people brought to near Tarbes.

30
Nephew and blood of the new saint come,
Through the surname he will sustain arches and roof:
They will be driven out put to death chased nude,
Into red and black will they convert their green.

31
The Holy Empire will come into Germany,
The Ishmaelites will find open places:
The asses will want also Carmania,
The supporters all covered by earth.

32
The great empire, everyone would be of it,
One will come to obtain it over the others:
But his realm and state will be of short duration,
Two years will he be able to maintain himself on the sea.

33
The cruel faction in the long robe
Will come to hide under the sharp daggers:
The Duke to seize Florence and the diphthong place,
Its discovery by immature ones and sycophants.

34
The Gaul who will hold the empire through war,
He will be betrayed by his minor brother–in–law:
He will be drawn by a fierce, prancing horse,
The brother will be hated for the deed for a long time

35
The younger son of the king flagrant in burning lust
To enjoy his first cousin:
Female attire in the Temple of Artemis,
Going to be murdered by the unknown one of Maine.

36
Upon the King of the stump speaking of wars,
The United Isle will hold him in contempt:
For several good years one gnawing and pillaging,
Through tyranny in the isle esteem changing.

37
The great assembly near the Lake of Bourget,
They will meet near Montmelian:
Going beyond the thoughtful ones will draw up a plan,
Chambery, Saint–Jean–de–Maurienne, Saint–Julien combat.

38
Sprightly love lays the siege not far,
The garrisons will be at the barbarian saint:
The Orsini and Adria will provide a guarantee for the Gauls,
For fear delivered by the army to the Grisons.

39
First son, widow, unfortunate marriage,
Without any children two Isles in discord:
Before eighteen, incompetent age,
For the other one the betrothal will take place while younger.

40
The young heir to the British realm,
Whom his dying father will have recommended:
The latter dead Lonole will dispute with him,
And from the son the realm demanded.

41
On the boundary of Caussade and Caylus,
Not at all far from the bottom of the valley:
Music from Villefranche to the sound of lutes,
Encompassed by cymbals and great stringing.

42
The humane realm of Anglican offspring,
It will cause its realm to hold to peace and union:
War half–captive in its enclosure,
For long will it cause them to maintain peace.

43
Too much good times, too much of royal goodness,
Ones made and unmade, quick, sudden, neglectful:
Lightly will he believe falsely of his loyal wife,
He put to death through his benevolence.

44
When a King will be against his people,
A native of Blois will subjugate the Ligurians,
Memel, Cordoba and the Dalmatians,
Of the seven then the shadow to the King, New Yearâs money and ghosts.

45
The shadow of the realm of Navarre untrue,
It will make his life one of fate unlawful:

The vow made in Cambrai wavering,
King Orleans will give a lawful wall.

46
In life, fate and death a sordid, unworthy man of gold,
He will not be a new Elector of Saxony:
From Brunswick he will send for a sign of love,
The false seducer delivering it to the people.

47
At the Garland lady of the town of Burgos,
They will impose for the treason committed:
The great prelate of Leon through Formande,
Undone by false pilgrims and ravishers.

48
Banners of the deepest part of Spain,
Coming out from the tip and ends of Europe:
Troubles passing near the bridge of Laigne,
Its great army will be routed by a band.

49
Garden of the world near the new city,
In the path of the hollow mountains:
It will be seized and plunged into the Tub,
Forced to drink waters poisoned by sulfur.

50
The Meuse by day in the land of Luxembourg,
It will find Saturn and three in the urn:
Mountain and plain, town, city and borough,
Flood in Lorraine, betrayed by the great urn.

51
Some of the lowest places of the land of Lorraine
Will be united with the Low Germans:
Through those of the see Picards, Normans, those of Main,
And they will be joined to the cantons.

52
At the place where the Lys and the Scheldt unite,
The nuptials will be arranged for a long time:
At the place in Antwerp where they carry the chaff,
Young old age wife undefiled.

53
The three concubines will fight each other for a long time,
The greatest one the least will remain to watch:
The great Selin will no longer be her patron,
She will call him fire shield white route.

54
She born in this world of a furtive concubine,
At two raised high by the sad news:
She will be taken captive by her enemies,
And brought to Malines and Brussels.

55
The unfortunate nuptials will be celebrated
In great joy but the end unhappy:
Husband and mother will slight the daughter-in-law,
The Apollo dead and the daughter-in-law more pitiful.

56
The royal prelate his bowing too low,
A great flow of blood will come out of his mouth:
The Anglican realm a realm pulled out of danger,
For long dead as a stump alive in Tunis.

57
The uplifted one will not know his scepter,
He will disgrace the young children of the greatest ones:
Never was there a more filthy and cruel being,
For their wives the king will banish them to death.

58
In the time of mourning the feline monarch
Will make war upon the young Macedonian:
Gaul to shake, the bark to be in jeopardy,
Marseilles to be tried in the West a talk.

59
Within Lyons twenty-five of one mind,
Five citizens, Germans, Bressans, Latins:
Under a noble one they will lead a long train,
And discovered by barks of mastiffs.

60
I weep for Nice, Monaco, Pisa, Genoa,
Savona, Siena, Capua, Modena, Malta:
For the above blood and sword for a New Year's gift,
Fire, the earth will tremble, water an unhappy reluctance.

61
Betta, Vienna, Emorte, Sopron,
They will want to deliver Pannonia to the Barbarians:
Enormous violence through pike and fire,
The conspirators discovered by a matron.

62
Near "Sorbia" to assail Hungary,
The herald of "Brudes" (dark ones?) will come to warn them:

Byzantine chief, Salona of Slavonia,
He will come to convert them to the law of the Arabs.

63
Cydonia, Ragusa, the city of St. Jerome,
With healing help to grow green again:
The King's son dead because of the death of two heroes,
Araby and Hungary will take the same course.

64
Weep Milan, weep Lucca and Florence,
As your great Duke climbs into the chariot:
The see to change it advances to near Venice,
When at Rome the Colonna will change.

65
O vast Rome, thy ruin approaches,
Not of thy walls, of thy blood and substance:
The one harsh in letters will make a very horrible notch,
Pointed steel driven into all up to the hilt.

66
The chief of London through the realm of America,
The Isle of Scotland will be tried by frost:
King and Reb will face an Antichrist so false,
That he will place them in the conflict all together.

67
A very mighty trembling in the month of May,
Saturn in Capricorn, Jupiter and Mercury in Taurus:
Venus also, Cancer, Mars in Virgo,
Hail will fall larger than an egg.

68
The army of the sea will stand before the city,
Then it will leave without making a long passage:
A great flock of citizens will be seized on land,
Fleet to return to seize it great robbery.

69
The shining deed of the old one exalted anew,
Through the South and Aquilon they will be very great:
Raised by his own sister great crowds,
Fleeing, murdered in the thicket of Ambellon.

70
Through an object the eye will swell very much,
Burning so much that the snow will fall:
The fields watered will come to shrink,
As the primate succumbs at Reggio.

71
The earth and air will freeze a very great sea,
When they will come to venerate Thursday:
That which will be never was it so fair,
From the four parts they will come to honor it.

72
The year 1999, seventh month,
From the sky will come a great King of Terror:
To bring back to life the great King of the Mongols,
Before and after Mars to reign by good luck.

73
The present time together with the past
Will be judged by the great Joker:
The world too late will be tired of him,
And through the clergy oath-taker disloyal.

74
The year of the great seventh number accomplished,
It will appear at the time of the games of slaughter:
Not far from the great millennial age,
When the buried will go out from their tombs.

75
Long awaited he will never return
In Europe, he will appear in Asia:
One of the league issued from the great Hermes,
And he will grow over all the Kings of the East.

76
The great Senate will ordain the triumph
For one who afterwards will be vanquished, driven out:
At the sound of the trumpet of his adherents there will be
Put up for sale their possessions, enemies expelled.

77
Thirty adherents of the order of Quirites
Banished, their possessions given their adversaries:
All their benefits will be taken as misdeeds,
Fleet dispersed, delivered to the Corsairs.

78
Sudden joy to sudden sadness,
It will occur at Rome for the graces embraced:
Grief, cries, tears, weeping, blood, excellent mirth,
Contrary bands surprised and trussed up.

79
The old roads will all be improved,
One will proceed on them to the modern Memphis:

The great Mercury of Hercules fleur-de-lis,
Causing to tremble lands, sea and country.

80
In the realm the great one of the great realm reigning,
Through force of arms the great gates of brass
He will cause to open, the King and Duke joining,
Fort demolished, ship to the bottom, day serene.

81
A treasure placed in a temple by Hesperian citizens,
Therein withdrawn to a secret place:
The hungry bonds to open the temple,
Retaken, ravished, a horrible prey in the midst.

82
Cries, weeping, tears will come with knives,
Seeming to flee, they will deliver a final attack,
Parks around to set up high platforms,
The living pushed back and murdered instantly.

83
The signal to give battle will not be given,
They will be obliged to go out of the park:
The banner around Ghent will be recognized,
Of him who will cause all his followers to be put to death.

84
The illegitimate girl so high, high, not low,
The late return will make the grieved ones contended:
The Reconciled One will not be without debates,
In employing and losing all his time.

85
The old tribune on the point of trembling,
He will be pressed not to deliver the captive:
The will, non-will, speaking the timid evil,
To deliver to his friends lawfully.

86
Like a griffin will come the King of Europe,
Accompanied by those of Aquilon:
He will lead a great troop of red ones and white ones,
And they will go against the King of Babylon.

87
A Great King will come to take port near Nice,
Thus the death of the great empire will be completed:
In Antibes will he place his heifer,
The plunder by sea all will vanish.

88
Foot and Horse at the second watch,
They will make an entry devastating all by sea:
Within the port of Marseilles he will enter,
Tears, cries, and blood, never times so bitter.

89
The walls will be converted from brick to marble,
Seven and fifty pacific years:
Joy to mortals, the aqueduct renewed,
Health, abundance of fruits, joy and mellifluous times.

90
A hundred times will the inhuman tyrant die,
In his place put one learned and mild,
The entire Senate will be under his hand,
He will be vexed by a rash scoundrel.

91
In the year 1609, Roman clergy,
At the beginning of the year you will hold an election:
Of one gray and black issued from Campania,
Never was there one so wicked as he.

92
Before his father the child will be killed,
The father afterwards between ropes of rushes:
The people of Geneva will have exerted themselves,
The chief lying in the middle like a log.

93
The new bark will take trips,
There and near by they will transfer the Empire:
Beaucaire, Arles will retain the hostages,
Near by, two columns of Porphyry found.

94
Scorn from N"mes, from Arles and Vienne,
Not to obey the Hesperian edict:
To the tormented to condemn the great one,
Six escaped in seraphic garb.

95
To the Spains will come a very powerful King,
By land and sea subjugating the South:
This evil will cause, lowering again the crescent,
Clipping the wings of those of Friday.

96
The Religion of the name of the seas will win out
Against the sect of the son of Adaluncatif:

Century X

The stubborn, lamented sect will be afraid
Of the two wounded by A and A.

97
Triremes full of captives of every age,
Good time for bad, the sweet for the bitter:
Prey to the Barbarians hasty they will be too soon,
Anxious to see the feather wail in the wind.

98
For the merry maid the bright splendor
Will shine no longer, for long will she be without salt:
With merchants, bullies, wolves odious,
All confusion universal monster.

99
The end of wolf, lion, ox and ass,
Timid deer they will be with mastiffs:
No longer will the sweet manna fall upon them,
More vigilance and watch for the mastiffs.

100
The great empire will be for England,
The all-powerful one for more than three hundred years:
Great forces to pass by sea and land,
The Lusitanians will not be satisfied thereby.

Almanacs: 1555–1563

Almanac of 1555

The soul touched from a distance by the divine spirit presages,
Trouble, famine, plague, war to hasten:
Water, droughts, land and sea stained with blood,
Peace, truce, prelates to be born, princes to die.

The Tyrrhenian Sea, the Ocean for the defense,
The great Neptune and his trident soldiers:
Provence secure because of the hand of the great Tende,
More Mars Narbonne the heroic de Villars.

The big bronze one which regulates the time of day,
Upon the death of the Tyrant it will be dismissed:
Tears, laments and cries, waters, ice bread does not give,
V.S.C. peace, the army will pass away.

Near Geneva terror will be great,
Through the counsel, that cannot fail:
The new King has his league prepare,
The young one dies, famine, fear will cause failure.

Writings of Nostradamus

O cruel Mars, how you should be feared,
More is the scythe with the silver conjoined:
Fleet, forces, water, wind of shadow to fear,
Sea and land in a truce. The friends has joined L.V.

For not having a guard you will be more offended,
The weak fort, Pinquiet uneasy and pacific:
They cry "famine," the people are oppressed,
The sea reddens, the Long one proud and iniquitous.

The five, six, fifteen, late and soon they remain,
The heirâs bloodline ended: the cities revolted:
The herald of peace twenty and three return,
The open-hearted five locked up, news invented.

At a distance, near the Aquarius, Saturn turns back,
That year great Mars will give a fire opposition,
Towards the North to the south the great proud female,
Florida in contemplation will hold the port.

Eight, fifteen, and five what disloyalty
The evil spy will come to be permitted:
Fire in the sky, lightning, fear, Papal terror,
The west trembles, pressing too hard the Salty wine.

Six, twelve, thirteen, twenty will speak to the Lady,
The older one by a woman will be corrupted:
Dijon, Guienne hail, lightning makes the first cut into it,
The insatiable one of blood and wine satisfied.

The sky to weep for him, made to do that!
The sea is being prepared, Hannibal to plan his ruse:
Denis [drops anchor], fleet delays, does not remain silent,
Has not known the secret, and by which you are amused!

Venus Neptune will pursue the enterprise,
Pensive one imprisoned, adversaries troubled:
Fleet in the Adriatic, cities towards the Thames,
The fourth clamor, by night, the reposing ones wounded.

The great one of the sky the cape will give,
Relief, Adriatic makes an offer to the port:
He who will be able will save himself from dangers,
By night the Great One wounded pursues.

The port protests too fraudulently and false,
The maw opened, condition of peace:
Rhone in crystal, water, snow, ice stained,
The death, death, wind, through rain the burden broken.

Almanac of 1558

Writings of Nostradamus

The young King makes a funeral wedding soon,
Holy one stirred up, feasts, of the said, Mars dormant:
Night tears they cry, they conduct the lady outside,
The arrest and peace broken on all sides.

Vain rumor within the Hierarchy,
Genoa to rebel: courses, offenses, tumults:
For the greater King will be the monarchy,
Election, conflict, covert burials.

Through discord in the absence to fail,
One suddenly will put him back on top:
Towards the North will be noises so loud,
Lesions, points to travel, above.

On the Tyrrhenian Sea, of different sail,
On the Ocean there will be diverse assaults:
Plague, poison, blood in the house of canvas,
Prefects, Legates stirred up to march high seas.

There where the faith was it will be broken,
The enemies will feed upon the enemies:
Fire rains [from the] Sky, it will burn, interrupted,
Enterprise by night. Chief will make quarrels.

War, thunder, forces fields, depopulated,
Terror and noise, assault on the frontier:
Great Great One fallen, pardon for the exiles,
Germans, Spaniards, by the sea the Barbarian banner.

The noise will be vain, the faltering ones bundled up,
The Shaven Ones captured: the all-powerful One elected:
The two Reds and four true crusaders to fail,
Rain troublesome to the powerful Monarch.

Rain, wind, forces, Barbarossa Hister, the Tyrrhenian Sea,
Vessels to pass Orkneys and beyond Gibraltar, grain and soldiers provided:
Retreats too well executed by Florence, Siena crossed,
The two will be dead, friendships joined.

Venus the beautiful will enter Florence.
The secret exiles will leave the place behind:
Many widows, they deplore the death of the Great One,
To remove from the realm, the Great Great one does not threaten.

Games, feasts, nuptials, dead Prelate of renown.
Noise, peace of truce while the enemy threatens:
Sea, land and sky noise, deed of the great Brennus,
Cries gold, silver, the enemy they ruin.

Almanac of 1560

Writings of Nostradamus

Dayâs journey, diet, interim, no council,
The year peace is being prepared, plague, schismatic famine:
Put outside inside, sky to change, domicile,
End of holiday, hierarchical revolt.

Diet to break up, the ancient sacred one to recover,
Under the two, fire through pardon to result:
Consecration without arms: the tall Red will want to have,
Peace of neglect, the Elected One, the Widower, to live.

To be made to appear elected with novelty,
Place of day-labor to go beyond the boundaries:
The feigned goodness to change to cruelty,
From the suspected place quickly will they all go out.

With the place chosen, the Shaved Ones will not be contented,
Led from Lake Geneva, unproven,
They will cause the old times to be renewed:
They will expose the frighten off the plot so well hatched.

Savoy peace will be broken,
The last hand will cause a strong levy:
The great conspirator will not be corrupted,
And the new alliance approved.

A long comet to wrong the Governor,
Hunger, burning fever, fire and reek of blood:
To all estates Jovial Ones in great honor,
Sedition by the Shaven ones, ignited.

Plague, famine, fire and ardor incessant,
Lightning, great hail, temple struck from the sky:
The Edict, arrest, and grievous law broken,
The chief inventor his people and himself snatched up.

Deprived will be the Shaven Ones of their arms,
It will augment their quarrel much:
Father Liber deceived lightning Albanians,
Sects will be gnawed to the marrow.

The modest request will be received,
They will be driven out and then restored on top:
The Great Great woman will be found content,
Blind ones, deaf ones will be put uppermost.

He will not be placed, the New Ones expelled,
Black king and the Great One will hold hard:
To have recourse to arms. Exiles expelled further,
To sing of victory, not free, consolation.

Almanacs: 1555–1563

The mourning left behind, supreme alliances,
Great Shaven One dead, refusal given at the entrance:
Upon return kindness to be in oblivion,
The death of the just one perpetrated at a banquet.

Almanac of 1562

Season of winter, good spring, sound, bad summer,
Pernicious autumn, dry, wheat rare:
Of wine enough, bad eyes, deeds, molested,
War, mutiny, seditious waste.

The hidden desire for the good will succeed,
Religion, peace, love and concord:
The nuptial song will not be completely in accord,
The high ones, who are low, and high, put to the rope.

For the Shaven Ones the Chief will not reach the end,
Edicts changed, the secret ones set at large:
Great One found dead, less of faith, low standing,
Dissimulated, shuddering, wounded in the boarâs lair.

Moved by Lion, near Lion he will undermine,
Taken, captive, pacified by a woman:
He will not hold as well as they will waver,
Placed unpassed, to oust the soul from rage.

From Lion he will come to arouse to move,
Vain discovery against infinite people:
Known by none the evil for the duty,
In the kitchen found dead and finished.

Nothing in accord, worse and more severe trouble,
As it was, land and sea to quieten:
All arrested, it will not be worth a double,
The iniquitous one will speak, Counsel of annihilation.

Portentous deed, horrible and incredible,
Typhoon will make the wicked ones move:
Those who then afterwards supported by the cable,
And the greater part exiled on the fields.

Right put on the throne come into France from the sky,
The whole world pacified by Virtue:
Much blood to scatter, sooner change to come,
By the birds, and by fire, and not by vers.

The colored ones, the Sacred malcontents,
Then suddenly through the happy Androgynes:
Of the great part to see, the time not come,
Several amongst them will make their soups weak.

They will be returned to their full power,
Conjoined at one point of the accord, not in accord:
All defied, more promised to the Shaven Ones,
Several amongst them outflanked in a band.

For the legate of terrestrial and dawn,
The great Cape will accommodate himself to all:
Tacit LORRAINE, to be listening,
He whose advice they will not want to agree with.

The enemy wind will impede the troop,
For the greatest one advance put in difficulty:
Wine with poison will be put in the cup,
To pass the great gun without horse-power.

Through crystal the enterprise is broken,
Games and feats, in LYONS to repose more:
No longer will he take his repast with the Great Ones,
Sudden catarrh, blessed water, to bathe him.

Almanacs: 1564–1567

Almanac of 1564

The sextile year rains, wheat to abound, hatreds,
Joy to men, Princes, King divorced:
Herd to perish, human mutations,
People oppressed and poison under the surface.

Times very diverse, discord discovered,
Council of war, change taken in, changed:
The Great Woman must not be, conspirators through water lost,
Great hostility, for the great one all steady.

The bit of the enemy's tongue approaches,
The Debonair one to peace will want to reduce:
The obstinate ones will want to lose the kinswoman,
Surprised, Captives, and suspects fury to injure.

Fathers and mothers dead of infinite sorrows,
Women in mourning, the pestilent she-monster:
The Great One to be no more, all the world to end,
Under peace, repose and every single one in opposition.

Princes and Christendom stirred up in debates,
Foreign nobles, Christ's See molested:
Become very evil, much good, mortal sight.
Death in the East, plague, famine, evil treaty.

Land to tremble, killed, wasteful, monster,

Captives without number, to do, undone, done:
To go over the sea misfortune will occur,
Proud against the proud evil done in disguise.

The unjust one lowered, they will molest him fiercely,
Hail, to flood, treasure, and engraved marble:
Chief of Persuasion people will kill to death,
And attached will be the blade to the tree.

Of what not evil? inexcusable result,
The fire not double, the Legate outside confused:
Against the worse wounded the fight will not be made,
The end of June the thread cut by firing.

Fine bonds enfeebled by accords,
Mars and Prelates united will not stop:
The great ones confused by gifts of mutilated bodies,
Dignified ones, undignified ones will seize the well endowed.

From good to the evil times will change,
The peace in the South, the expectation of the Greatest Ones:
The Great Ones grieving Louis too much more will stumble,
Well–known Shaven Ones have neither power not understanding.

This is the month for evils so many as to be doubled,
Deaths, plague to drain all, famine, to quarrel:
Those of the reverse of exile will come to note,
Great Ones, secrets, deaths, not to censure.

Through death, death to bite, counsel, robbery, pestiferous,
They will not dare to attack the Marines:
Deucalion a final trouble to make,
Few young people: half–dead to give a start.

Dead through spite he will cause the others to shine,
And in an exalted place some great evils to occur:
Sad concepts will come to harm each one,
Temporal dignified, the Mass to succeed.

Almanac of 1566

For the greatest ones death, loss of honor and violence,
Professors of the faith, their estate and their sect:
For the two great Churches diverse noise, decadence,
Evil neighbors quarreling serfs of the Church without a head.

Waste, great loss, and not without violence,
All those of the faith, more for religion,
The Greatest Ones will lose their lives, their honor and fortunes
Both the two Churches, the sin in their faction.

Writings of Nostradamus

For the two very Great Ones pernicious loss to arise,
The Greatest Ones will cause loss, goods, of honor, and of life,
As much great noises will run, the urn very odious,
Great maladies to be, meeting–house, mass in envy.

The servants of the Churches will betray their Lords,
Of other Lords also by the undivided of the fields:
Neighbors of meeting–house and mass will quarrel amongst them,
Rumors, noises to augment, to death are several lying.

Of all blessings abundance, the earth will produce for us,
No din of war in France, sedition put outside:
Man–slayers, robbers one will find on the highway,
Little faith, burning fever, people in commotion.

Between people discord, brutal enmity,
War, death of great Princes, several parts:
Universal plague, stronger in the West,
Times good and full, but very dry and exhausted.

The grains not to be plentiful, in all other fruits, plenty,
The Summer, spring humid, winter long, snow, ice:
The East in arms, France reinforces herself,
Death of beasts much honey, the place to be besieged.

Through pestilence and fire fruits of trees will perish,
Signs of oil to abound. Father Denis not scarce:
Some great ones to die, but few foreigners will sally forth in attack,
Offense, Barbarian marines, and dangers at the frontiers.

Rains very excessive, and of blessings abundance,
The cattle price to be just, women outside of danger:
Hail, rain, thunder: people depressed in France,
Through death they will work, death to reprove people.

Arms, plagues to cease, death of the seditious ones,
Great Father Liber will not much abound:
Evil ones will be seized by more malicious ones,
France more than ever victorious will triumph.

Up to this month the great drought will endure,
For Italy and Provence all fruits to half:
The Great One less of enemies prisoner of their band,
For the scroungers, Pirates, and the enemy to die.

The enemy so much to be feared to retire into Thrace,
Leaving cries, howls, and pillage desolated:
To leave noise on sea and land, religion murdered,
Jovial Ones put on the road, every sect to become angry.